Praise for

What Sheep Think About the Weather

"Blend the lively, buoyant writing of an award-winning nonscientist with a boatload of direct experiences with animals, and you have this book. I found it both illuminating and entertaining from start to finish."

—Jonathan Balcombe, *New York Times* bestselling author of *Super Fly* and *What a Fish Knows*

"Amelia Thomas is a powerful storyteller and has hit it out of the park in this contemporary anthology of our relationship with animals. From cuttlefish to cows, this brilliantly written and emotionally charged page-turner is a masterpiece and masterclass for all animals—human and nonhuman."

—Pilley Bianchi, author of *For the Love of Dog: The Ultimate Relationship Guide* and co-teacher of Chaser, known as "the smartest dog in the world"

"A must-read for anyone interested in how animals talk with one another and with us. Thomas steers us from our human-centric view of other animals—domestic and wild, big and small. This book will open all your senses. Awe-inspiring, important, and timely."

—Marc Bekoff, PhD, author of *The Emotional Lives of Animals: A Leading Scientist Explores Animal Joy, Sorrow, and Empathy—and Why They Matter*

"Every single page of this marvelous book will challenge you, even enchant you. With intellectual thrills in spades paired with Amelia's conversational style, it's a delight to read. I cannot recommend it enough!"

—Jeffrey Moussaieff Masson, author of *When Elephants Weep: The Emotional Lives of Animals*

What Sheep Think About the Weather

How to Listen to
**WHAT ANIMALS ARE
TRYING TO SAY**

Amelia Thomas

sourcebooks

Copyright © 2025 by Amelia Thomas
Cover and internal design © 2025 by Sourcebooks
Cover design by Pete Garceau
Cover images © DonNichols/Getty Images, Eric Isselee/
Shutterstock, wacomka/Shutterstock
Internal design by Laura Boren/Sourcebooks

Published by Sourcebooks
1935 Brookdale RD, Naperville, IL 60563-2773
(630) 961-3900
sourcebooks.com

Cataloging-in-Publication Data is on file with the Library of Congress.

Printed and bound in the United States of America.
POD

For all the silenced voices

Introduction

To Listen Like Simona

> "The destiny of Man is to unite, not to divide."
>
> —T. H. WHITE, *The Once and Future King*
>
> "The method, then, is simply to go as close to the frontier as possible and peer over it with whatever instruments are available."
>
> —CHARLES FOSTER, *Being a Beast*

AUGUST

It's actually mine.

I am sitting barefoot in the sunshine on my new front doorstep, cradling my morning coffee and trying hard to believe that this old farm, this rambling place of listing floors and storm-weathered shingle, now *really belongs to me.*

To my left spreads its orchard of arthritic apple trees. To my right its five-acre wood, home to a loquacious cock pheasant whom Indy, our eleven-year-old, has named Siegfried, along with Fräulein and Hildegard, his two-hen harem. Ring-necked pheasants originated in Central Asia, and Siegfried, as he struts and shimmies, is the very model of a pompous, nervous, screeching grand vizier. Out front, a flock of starlings roosts in one old oak, past which puff antique tractors and elderly bicycle tourists. Behind me, a tumbledown hay barn, love nest of a rock pigeon couple, who, unlike our sexed-up starlings, mate for life; and beyond it, a marshy, viridian meadow that slopes down to the oxbow of a tidal river. Bald eagles, ospreys, and red-tailed hawks patrol the river's fringes. Polled Herefords and Jerseys moo from the opposite bank. Cormorants festoon themselves like drying laundry on a long-dead conifer; turkey vultures ride the thermals overhead. At dusk, spring peepers chorus from the meadow's soggy bottom; at night, coyotes scream from the wood.

"It's mine," I close my eyes and whisper. "It really is."

"BRRRR-ZZZZZZZZZZ-UMMMM."

A violent aerial assault startles me out of my reverie. I jump, spill my coffee into my lap, and squeal—an involuntary response from deep within my limbic system. "Hornet!" my lizard brain shrieks. "Retreat, retreat!"

"BRRRR-ZZZZZZZZZZ-UMMMM."

With a flash of emerald feather, my prefrontal cortex catches up. This is no hornet, but a ruby-throated hummingbird, *Archilochus colubris*, a brilliant, lucent lozenge about the size of my thumb, with wings that beat sixty times per second and a heart that drums an unimaginable twelve hundred times a minute. He zips down, turns a tight aerial doughnut, and hovers several inches from my face, eye to eye, nose to rapier beak.

Hummingbirds are intensely territorial. These rubythroats, with their verdigris plumage and scarlet mantles, migrate some two thousand five hundred miles, five hundred of these over perilous open ocean, south to their wintering grounds between Yucatán and Panama, before, astonishingly, returning each spring to hatch young in nests bound by spiders' webs, from eggs less than half the size of your pinkie fingernail. Little wonder, after such an odyssey, that he's quite particular about who moves into the neighborhood.

He issues a series of quick, angry-sounding squeaks. *Actually*, this tiny, lion-hearted emperor seems to be saying, *this is MINE*. He zooms off in a circle and returns for a second offensive, then a third.

"All right, all right." I rise and drain what's left of my coffee. "I understand."

Apparently satisfied, he buzzes off to eye me beadily from the topmost cane of the raspberry patch.

But was that, I wonder as I go inside, *really* what he was saying?

I wish I knew for certain.

Over the course of my life, I have wished this repeatedly. I wished it at age thirteen, when I brought home an oily pigeon I found hunched in a Birmingham alley as commuters and shoppers marched by. "Is it your wing? Your foot? Have you lost your flight feathers?" I asked as I transported him to safety on the number fifty-one bus. I wished it in my early twenties, when Sherpa, my rescued Afghan hound, tore to shreds just one of my future husband's four Beatles cushions (sorry, Ringo). I wished it as a newlywed in Amsterdam, each time Charlie, our pet wallaby, looked me straight in the eye and took a defiant bite of stolen Spanish onion, though zoological wisdom held that wallabies would eat only timothy hay and salad vegetables. "I wish you could tell me where it hurts," I crooned years later to Ivy, our Plymouth Rock chicken, as she lay dying in a blanket-lined box beside the living-room fire.

Now here I am, on my own farm, with hummingbirds talking to me on what I think they think is their front doorstep, and I remain none the wiser. How can we humans better understand what animals great and small (or both, in this mighty miniature's case) are trying to tell us?

I enter the kitchen, where Dolly, our 120-pound Daniff, and Puff, our white bull terrier ("One part sugar, two parts spice," her mother's owner warned me), are asleep in a tangle of limbs, and glance out of the window to where Constance and Agnes, our brand-new three-month-old Kunekune piglets, should be snoozing in their brand-new pen.

Only they aren't.

Instead, there goes Agnes through the sweet-pea patch, free as a bird and with Constance in hot pursuit.

My first day as a farmer, and I'm already failing.

I ought to be better at this.

For five documented centuries, but likely many more, my forebears were farmers.

As a child, I visited elderly relatives in their farmhouses on the whaleback brink of England, where it cedes into Wales. In one lived Great-Aunt Nora, deaf and quite batty, who regularly tossed the good china into the duck pond. In another, a dark, *Cold Comfort*-ish place, dwelled one pair of ancient maiden cousins and another of barren peahens, both in mourning for their recently departed cock. One of my earliest memories is of being lifted onto the roof of a Land Rover while my grandpa and Great-Uncle Joe rallied to gather a flock of escaped Welsh Mountain sheep. Growing up, I could think of no better life than one lived beside bovines and ovines in a cold stone farmhouse, with mud under your fingernails and wisps of hay in your hair and barns and byres to muck out on frigid winter mornings.

Who, I wondered, wouldn't want that sort of idyll?

The answer, to my dismay, was my parents, who decamped to suburbia. I wanted wild strides across the Black Mountains with my pack of hoary sheepdogs. Instead, I got standing at the back door with a tin of Friskies, calling, "O'Malley! Duchess! Here, kitty-kitty-kitty!"

My childhood, then, was spent with homelier creatures: with rescued rabbits and guinea pigs; the injured pigeon I named Peter, who surprised me by laying an egg in my underwear drawer; a half dozen mice and several rats, all otherwise destined for snake food; the nine or ten goldfish I won at the fair. One spring, a vixen brought her cubs to play in the dappled sunlight of our back garden. On summer evenings, I fed the cats' Friskies to visiting hedgehogs.

At age ten, my first bit of journalism appeared in the local newspaper. "Heartless Dumping Threatens Wildlife," read the caption. "I am writing to tell you about people dumping rubbish in the canal…" Thirteen years later, the day I got my first job as a reporter, I marched promptly to the pound and returned with a scruffy, schnauzery mongrel I named Milo. Milo, like Auntie Nora, was loopy. He chased police horses and recumbent cyclists and electric wheelchairs. He once launched himself through my second-floor window because he saw a dog on the street below chasing a ball; he bounced off the hood of a parked car, hit the ground, and stole the ball away. I had no idea why he did the things he did or, most of the time, what it could possibly mean. But I kept him through moves and marriage and small children. When he died, aged about sixteen, he'd lived on three continents in twelve houses and had gently welcomed all five babies into our family.

Other animals followed: the Beatle-biased Afghan, two dachshunds, and a solid-citizen French mastiff. A Siamese fighting fish, *Betta splendens*, abandoned after being part of an art installation. A three-legged kitten from the streets of Jerusalem; a wallaby purchased by accident from a Dutch classifieds ad; a budgie from a Russian circus; a series of rescued chickens. From a decrepit pet shop I passed on my way to work, I bought and rehomed all the animals in direst need, except for the glum-looking axolotl marked "Not For Sale," who rose and fell perpetually in his tank.

Elsewhere, in pursuit of stories, I met grander animals. A winsome giraffe. Pampered tigers. Sanctuary elephants in Laos; working elephants in India. Camels, Bactrian and dromedary, placid and irascible. In time, I acquired a big red horse—named Major, but labeled "difficult"—and purchased, sight unseen, a beautiful blond Icelandic pony called Andvari, whom his seller described as "sensitive."* The

---

* The difference between a horse and a pony is purely one of height: under 14.2 hands high at the withers is considered a pony; 14.2 hands high and above is a horse. Nevertheless, the first time I flicked a fly from his forelock, he performed a vertical cartoon-character leap and bolted five hundred yards.

And then, after many moves and many countries and much yearning, it finally happened. We bought our own farm: not cold, not stone, not on the cusp of Wales, but very old and rambling and in Nova Scotia, Canada. A place from which Gal, my Finnish husband, and I could work remotely; in which Indio, Zeyah, and Cairo, our three youngest children, could be the country kids I'd so longed to be; and to which Cassidy and Tyger, our two oldest, could return during university holidays.

This is it, my DNA seemed to say: *home*. Where I could mourn my own peacock, or toss china into my very own pond, or mislay my own piglets after barely twelve hours in residence. Because though I had always anticipated some ancient, atavistic knowledge would kick in the moment I officially became a farmer, ancestral wisdom, it seems, did not get the memo. I still feel that desperate urge to vault the species divide to find out what animals are saying and the inkling that if only I can, then rich conversations await.

But to converse with a piglet, I'll have to catch one first.

"Why are you out? Where are you going? Come back! Come back!"

Dropping my coffee cup, I race out to the garden through which two gingersnap piglets are currently hurtling, headed, like a long-gone flock of Welsh Mountain sheep, for the wide blue yonder.

Neither listens: not Constance, a chunky redhead, nor Agnes, her svelter, feistier sister, with wattles and spots and one black leg as if

---

storied heritage of Iceland's only equine (Icelandic law forbids the importation of any other breed) demands that Icelandics, though rarely tall enough even on tiptoes, be called horses. This suggests there is something inglorious, something *lesser*, about a pony. Anyone who has ever ridden one, however, knows that big hearts and strong opinions come in small, hairy packages.

somebody dipped her into an inkpot. We brought both home late last night, fat bottoms squeezed into dog crates, after receiving a 7:00 p.m. text message from their seller: URGENT Pls pick up piggies 2nite Must Leave 4 Wkend.

They emerged from their carriers blearily, blinked their little piggy eyes, and promptly fell asleep in the deep, fragrant hay of the A-frame "pigloo" the children built for them. I tiptoed away.

Pigs are reasonable, I thought. *Easy.*

"Sweet dreams, little ones."

I fed them breakfast at 7:00 a.m., relieved to see they'd made it, coyote-proof, through the night.

Yet here they are now, heading for calamity of their own free will.

Panicked, I toss grain into a bucket and jangle it. They U-turn, stubby legs pinwheeling, and careen back toward me. "Come on, then! Back inside! Good girls!"

From down in the meadow, I hear the horses, who until yesterday were perfect strangers, squealing. Is something attacking them? Are they attacking each other? Or is one of them stuck in something? Horses, it seems, are forever looking for creative ways to die. This hubbub is accompanied from the house by frenzied barking: Dolly is at the window, and, having seen the piglet derby in progress, is making her feelings known. But what is she saying? Her parents are livestock guardians, keeping their own farm safe from bears and wolves and cougars. Is she telling me she can help? Is she guarding me, her livestock, from them? Or is she simply saying they look like a tasty snack, and could I please oblige by opening the back door? Puff joins in. Then, in their ebullience, they set upon each other. Teeth gnash. Fur flies. I try to ignore the blue-ribbon dogfight and focus on the piglets.

"This way, girls!"

They race behind me, chasing the bucket, and with some

maneuvering of porcine body parts, I close their apparently useless gate behind us.

"All right. Now, are you going to…"

Before I can finish, Agnes is off again, tiny snout levering up the nice, neat, ineffectual fence, little wriggly body close behind. Constance, barrel-shaped, follows suit. This time, they head for the road.

Horrified, I give chase. How am I to explain to eleven-year-old Indy, currently at the library selecting summer reading, that his new pets met a grisly doom before he made it to Nonfiction? I wrangle them in. Gleefully, they let themselves back out. I strengthen the fencing. They push harder, straight underneath it. I shove them, undignified and unladylike, back into last night's dog crates. Constance goes in face-first, bottom bulging through the bars.

Then off I sprint to check on the horses, who treat the arrival of this sweaty maniac with understandable circumspection. I receive a tooth-jarring jolt for forgetting not to touch their newly electrified fence. Hastening back to great oinks of disgruntlement, I haul stray planks and posts, pile everything along the piglets' nonelectric fence line like some crazed game of Jenga, and nail bits together until there's absolutely, positively no way out.

I release them from their carriers and dissolve to the ground, panting, in this shoddy new Porkatraz.

Delighted, they dissolve with me.

Within seconds, both are dozing contentedly, Agnes with her head beneath my knee and one hind foot on her sister's snout. Constance purrs and squeezes her eyes shut as I stroke her ears, her tiny cloven hooves. She shifts her belly toward me. *Rub here*, she seems to be saying. Not daring to leave the piglets unattended, I oblige.

Inside the house, the dogs simmer down. The corners of Constance's mouth, I notice, are upturned into a little smile. What were the piglets telling me, I wonder, when they attempted their great escape?

That question, yet again.

I pull out my phone and text their seller.

Are they missing their mother? I hazard.

LOLZ not likely, my phone buzzes a reply. Their moms bn sick of them 4 weeks haha lil buggers.

"I know how she feels," I inform the piglets.

They decline to reply.

It wasn't supposed to have happened like this, I reflect, as I mop my brow and regain some semblance of composure, the animals arriving two by two like applicants to the ark. We'd intended to stagger the newcomers, to give everyone a chance to settle in: dogs and cat one week, two piglets the next, one horse the week after, and the other a week after that. But then, like British buses, they had all come at once.

First Major, my "difficult" big red horse, whom I'd boarded several hours away since moving him four thousand miles cross-country last year. He was deposited lame and skinny, hooves abscessed, covered in bite marks, and unable to navigate the meadow's ruts. I watched this once powerful horse, on whom I'd galloped, devil-may-care, through West Coast forests, stumble his way to the water trough. I could count his ribs. Taut muscles pronged his neck. Gone was his big, fat, pattable chestnut bum.

I wish you could tell me what happened, I thought and felt terrible.

I had barely an hour to lament his condition, however, before a zealous friend delivered Icelandic Andvari, whom he'd cared for at his own farm, three weeks early.

"There is no more sagacious animal than the Icelandic horse," wrote Jules Verne in *Journey to the Center of the Earth*, our oldest child, Cassidy's, favorite book. "He is stopped by neither snow nor storm, nor impassable roads, nor rocks, glaciers, or anything. He is courageous, sober, and sure-footed. He never makes a false step, never shies."

Verne evidently never met an Andvari, who arrived terrified of his new hay net: a bag comprising 5 percent green string and 95 percent fresh air.

And then came the piglets.

Now here I sit, with both snoring under knee, awaiting someone, anyone, to come home and help me.

And while I do, I think of Simona Kossak.

Several years ago, a friend sent me the most tantalizing black-and-white photograph. It depicts a wild boar, about the size of a Highland heifer, nibbling crumbs from a dining table made from the cross section of an enormous, primeval tree. The room is stuffed with oil lamps and carriage clocks. A candelabra burns in one corner. In front, a young woman with bangs and braids and a woolen cardigan, half Greta Thunberg, half Pippi Longstocking, looks on with pride and adoration.

"A common meal," reads the caption, "in the company of a particular household member—Simona with the sow Żabka."

"This could be you," my friend wrote, signing off with two emojis: a smiley face and a little pink pig.

Immediately, I was intrigued. Who was this woman, and why were her life choices so much better than mine?

I looked her up.

Her name was Simona Kossak, and she'd lived in an ancient forester's lodge called Dziedzinka, deep within the Bialowieza forest, about 150 miles northeast of Warsaw, with a lynx named Agatka, a raven named Korasek, Żabka the boar, and her partner, Polish wildlife photographer Lech Wilczek.

Little has been written about her in English, and my Polish begins and ends at "Proszę dwie wiśniówki" ("Two cherry vodkas, please"). Her one book with an English translation is long out of print. But I dug and collected and pieced together all I could. Simona, I learned, was born in 1943 to a prominent Krakow family; her father, grandfather, and great-grandfather were lauded artists, all specializing in the gun smoke and eye whites of cavalry battles. Expectations ran high. But as a girl with a penchant for animals, not a boy with wild artistic talent, Simona proved a disappointment and was largely ignored. Her childhood's saving graces were her father's Great Danes, and her bedroom sick bay full of rescued birds and hedgehogs. "My entire sympathy for the animal world," she said later in life, "came from my family home."

Simona trained as a zoologist. Her first job offer came from the Mammal Research Institute in the remote, wolf-laden Bialowieza National Park, where, in its depths, she found the ruins of a woodcutter's lodge surrounded by herds of woodland elk and bison. She moved in, rebuilt, and populated her new home with the sorts of relationships she found effortless. A donkey named Hepunia (who, a 1970s Polish magazine article recounts, "tried to flee to the Soviet Union twice"). Twin elk calves named Pepsi and Cola. A hedgehog, a black stork with its nest in a trunk in her bedroom, the lynx and a dachshund who both slept in her bed; dormice, peacocks, rats, and buzzards; a long-eared owl with injured legs; and a tawny with a broken wing. Żabka the boar lived with her for eighteen years. Korasek, her raven, terrorized nearby villagers, who thought Simona a witch even after she received a doctorate, in 1980, in zoopsychology.

Simona lived in that forest, with Lech and their myriad animals, as a professor of forest sciences and, eventually, national director of the Department of Forests until her early death in 2007. She listened closely to both her own animals and those of the natural world around her. She filmed nature films about them, wrote books, and authored dozens of scientific papers.

What, I have wondered repeatedly since I first saw that photograph, did Simona know about listening that I don't? How could she have such effortless friendships with so many kinds of animals? It was she who'd inspired me to adopt Constance and Agnes (uncertain how the family would feel about a boar, I could, I felt, get away with piglets), and to dream of a farm where I could finally find out what animals were trying to say.

Whatever she did, I decide, as two snouts nudge my bottom and the horses start screaming again in the field, I want to do it too.

A crunch of tires on gravel signals the return of the cavalry.

I sit waiting, piglets dozing, until I'm found, so quietly that Dolores, our resident groundhog, emerges from her hole and sits munching ruminatively on day lilies.

Then among Gal, Indy, and me, we secure every inch of the pigpen. I watch how Agnes patrols the perimeter. She gives the posts a good rattle, nudges the wire. It seems she knows exactly where the fence's weak points were. I have the feeling she'll be performing the same sort of forensic investigation on me.

Indy shows me his library books—*The Horse and his Boy* and *Five Children and It*—and outlines his plans for September's back-to-school talent show: a mime routine, he says, in full Pierrot black-and-white.

We walk the dogs (uninjured). We check on the horses (unstuck). I head upstairs to change.

Books, I consider as I climb out of my piggy dungarees, are full of animals talking to people. Creatures speak in the Mahabharata, the Qur'an, the Jatakas, the Bible, and abundantly throughout children's literature. My favorites, as a child, were White's *The Once and Future King*, wherein Merlyn educated the future King Arthur by transforming him into a perch, a hawk, an ant, an owl, a goose, and a badger; and Kipling's *Just So Stories*:

> There was never a king like Solomon,
>
> Not since the world began;
>
> But Solomon talked to a butterfly
>
> As a man would talk to a man.

But what about me? I'm not expecting Arthurian- or Solomon-level communication, but how can I better understand the animals in my life?

I sit down on the bed and leap straight up again. Winnie, our Himalayan cat, a five-pound, grumpy-faced fluff ball with a ridiculous "lion cut" hairstyle, has peed, lavishly and equitably, at the very center of the duvet.

"Why, Winnie?" I demand. "*Why?*"

Winnie stares phlegmatically from the doorway, then sashays, tail tip flicking, away.

Later, bedsheets bleached and hung out to dry, I settle into a deck chair with James Herriot's veterinary memoir, *If Only They Could Talk*.

In the orchard, Indy practices his talent-show routine. Gravely he climbs a ladder, blows up a balloon, gets trapped in a locked glass box. The ruby-throated hummingbird spectates suspiciously from the raspberries, the hop-along horse with interest from the meadow gate.

But they can *talk, can't they?* insists a little voice inside me. *You just don't know how to listen.*

Did Simona Kossak have mornings like this, I wonder, with her boar and her elk and her donkey? Did her lynx ever pee on her bed? What would she do if she didn't understand what her animal companions were trying to tell her?

Simple, the voice replies.

She'd experiment.

She would learn.

She would try to find out.

And then she might write a book.

Can animals talk? Few would dispute these days that at the very least, they communicate with one another: A wittering flock of starlings is not wittering mindlessly en masse; a mother duck calling her duck-lings isn't making meaningless sounds on a purely instinctual urge. Bees exchange directions to burgeoning forage grounds via "waggle dances." Elephants express love and grief through body language, sound, and infrasound. Dolphins each possess a "signature whistle" that functions like a human name. Mother cows and their calves, along with ewes and their lambs, have individualized calls for one another. Bats, whales, and chickadees all possess local dialects, as do fruit flies, who can even learn one another's if they live together. "Remarkably," says a 2018 scientific paper titled "Drosophila Species Learn Dialects Through Communal Living," "partial communication between some species is enhanced after a cohabitation period that requires exchange of visual and olfactory signals."

But can they talk to *us*? I've never doubted that either. Our dogs wag their tails at us when they decide it's time for a walk. Our cats meow to be let in, then out, then straight back in again, and then pee, with precision, on our beds. Your pet rabbit or budgie might solicit attention, or tell you to piss off, with subtle body language. Our chickens told us, both vocally and physically, when they wanted treats or to be let out of their coop to forage in the flower beds. No human words are spoken in any of these instances. Still, to me, it's all talking.

Most of us, too, can instinctively understand unfamiliar animals' signals. You sense when a strange dog is behaving threateningly: those raised hackles, that slinking posture, that low growl. You know, a fragment of a second before they flee, that a wild bird or deer or squirrel thinks you've come too close. You also know what isn't true communication: that the puffer fish you meet while snorkeling isn't really smiling at you, and the grin of a manta ray is a matter of physiology, not pleasure at our presence. We have a hunch when we're anthropomorphizing, but can also accuse ourselves of doing so when we're simply looking more deeply into the animal world—or overestimating the uniqueness of the human one. "This is going to sound crazy," a beekeeping friend told me recently, "but when I got back from vacation, one flew onto my chest as if welcoming me home." You're not crazy, I replied: Several scientific studies have documented bees' ability to recognize human faces.

Yet beyond the things we already know about animals' communications, how much deeper can we go? What more do they say to us that we're missing, and how can we learn to understand? Can we find a chink in that frontier between *Them* and *Us*, and which tools can we employ to chip away at it?

In some ways, animals are simpler than humans. Hamsters don't deliberately confound or obfuscate. Donkeys don't gossip. An iguana will not gaslight you. Animals say what they mean. Yet that's not to say this content is clear, or that we're always aware it even exists at all.

And so, sitting here, watching a hummingbird watch an eleven-year-old ride an imaginary horse while a real-life horse stands on three legs and looks curiously on, I decide to write this book.

Only: *how?*

"Primary sources, girls! Primary sources!" I hear, in my mind's ear, the rallying cry of Mrs. Jenny Johnson, my high school history teacher, which stayed with me throughout my years on the newspaper beat.

I shall go, I decide, straight to the horses' mouths: to the experts and, better yet, to the creatures themselves. What are animals telling us? What are animal scientists discovering? What do animal trainers know? What do those in the intuitive realms—the trackers, the animal communicators—feel?

I'll spend my first year as a farmer listening to all those listeners: to experts whose research I've pored over and whose courses I've taken, to those who've devoted their lives to listening to what animals have to say to us humans. And I'll engage the help of my own motley crew, more commonplace than Simona Kossak's, to help me understand. Dolly, our Daniff—half Great Dane, half Presa Canario mastiff—who adores her family and is suspicious of all strangers. Puff, our bull's-eyed bull terrier, who loves people so much that she'd like to climb into their pockets. Winnie, the selectively incontinent Himalayan. Andvari, the equine Viking who jumps if someone sneezes, and Major, the skinny big red horse, whom former owners have described variously as "quirky," "unusual," and "a bit of a dick." And, of course, Constance and Agnes, the Incredible Escaping Piglets.

In the orchard, Indy wraps up his routine. From my deck chair, I give him a big thumbs-up. This won't be a book, I resolve, about how animals communicate with one another: I've read many brilliant ones on this already, some by specialists I intend to talk with for my own. Nor will it be about how to talk *to* your animal, though how and when we talk will certainly come into it. I remember a quote from Democritus. "It is greed," he said, "to do all the talking, but not be prepared to listen." This book will be about how, why, and what it means to listen. With the guidance of the world's best animal listeners, I'll endeavor to slip through that chink in the great frontier that divides our species from the planet's roughly nine million others and have a chat on the other side.

Later, I email Anna Kaminska, Simona Kossak's Polish biographer, and bombard her with questions. How, I want to know, did Simona make listening to animals look so easy? Despite the language barrier, I get the message: Simona was an animal scientist, Anna tells me, and an animal trainer, and she used her intuition. She was also an exceptionally good listener. Simona, I decide, will be my barometer; if I can listen even half as well as she could, I'll know I've achieved my goal. For inspiration, I print out that first photograph of Simona and Żabka, her gargantuan boar, and pin it to the kitchen wall beside the piglet-vista window.

Most of our daily interactions with other species involve us talking, not listening. "Sit" and "stay" are our mainstays: We tell animals what to do, what not to, and when to do or not do it. We teach them to understand the handful of human words we feel they need to know. Sometimes, intentionally or otherwise, we also teach them to imitate us. Take the eight foul-mouthed parrots at the Lincolnshire Wildlife Park whose penchant for expletives prompted staff to erect a "Warning" sign for visitors, or Hoover the orphaned Maine harbor seal, who learned to copy the voice of his surrogate parent, New Englander George Swallow, thereafter delighting '80s TV audiences by shouting, "Hello, there!" and "Come over here!" with a distinct Boston twang.

But how often, if ever, do we stop talking and *truly, deeply* listen?

I am not an expert in this domain. I am an amateur naturalist: I know lots of fairly useless facts about the natural world, including several about blaspheming birds, and I seize every opportunity to watch it closely. I'm also an amateur trainer. I teach my dogs silly tricks that make me smile and win them treats, and I train hand-me-down horses, seeking holes in human-horse communication to fill with mutual understanding. I am definitely not a scientist and have no special intuitive abilities. I have never, unlike some of the exceptional people we'll be meeting, taught animals American Sign Language or human speech to describe objects or concepts or their feelings.

But I *am* a decent listener. For work, I've spent a lot of time listening, to clowns and gurus and terrorists, to shopkeepers and dominatrices and geriatric drug smugglers, and found scraps of commonality every-where. I love getting wrapped up in other people's thoughts and ideas, trying to imagine how they experience the world we share, often so differently from how I do. At home, I've listened to our five children from toddlerhood through teenagerdom, none of whom have ever been short of a thing or two to say. It feels natural to want to include animals too.

This book, then, is an exploration, a quest to find out what might happen when we invite the animal kingdom to really talk to us. Which techniques and tidbits can we all learn to listen better in a multitude of contexts: in the science lab, in a training session, in a wood or a city, on a beach, or out standing in your field?

What I present is my own subjective experience, delivered to you through my biased choice of words, to try to convey what happened, as I alone see it, when I spent a year learning to listen to animals (as Thoreau said in *Walden*, "I should not talk so much about myself if there were anybody else whom I knew as well"). It is written with a keen awareness of my human limitations: my five puny senses, my Western-civilization emphasis on brain over body, of thinking over feeling. Someone might accuse me of anthropomorphizing. Someone else might accuse them of being anthropocentrists, in anthropodenial.[†] But I have tried to keep my mind open and my biases in check: that science is hard, and training is complicated, and animal communication is the stuff of crystal balls and wishful thinking. Still, ask my dogs next summer, and they might tell you a completely different story. Ask my pigs, and they might shrug, if they had shoulders, and reply, "Who cares?" Ask my platinum blond pony, and he might jump at your shadow, then head for the hills.

---

†   The late, great ethologist and primatologist Frans de Waal (1948–2024) distinguished between anthropomorphism (ascribing human-like qualities to animals), anthropocentrism (the notion that humans are at the center of the universe and animals exist for our enjoyment), and anthropodenialism (refusal to see human-like qualities in animals and vice versa).

PART I

The Thinkers

The Science of Listening to Animals

sci · ence /ˈsīəns/

noun

the systematic study of the structure and behavior of the physical and natural world through observation, experimentation, and the testing of theories against the evidence obtained.

1

I Say Talking, and You Say Speaking

The Semantics of Saying Something

"The one great barrier between brute and man is Language. Man speaks, and no brute has ever uttered a word. Language is our Rubicon, and no brute will dare to cross it."

—FRIEDRICH MAX MÜLLER, linguist, 1862

"The difference in mind between man and the higher animals, great as it is, is certainly one of degree and not of kind."

—CHARLES DARWIN, naturalist, 1871

SEPTEMBER

I am sitting on the floor of our darkened farm workshop with an eleven-year-old child and a big green plastic tub.

"Ready... Set... Go!"

Simultaneously, Indy and I switch on our flashlights, and he's off, with a commentary better suited to a horse race than a tubful of startled earwigs.

"And it's Mabel... She's running for the shelter. She's almost there... Watch her, Mummy, watch her... There she goes, straight underneath it. Oh, but Cecilia—no, no, it's Gwendolyn—is catching her up. And look at Rita: I *think* it's Rita. She's staying put. She couldn't care less. Write that down, Mummy—quick, quick—'Rita couldn't care less.'"

With clipboard and pen, notes are scribbled, times recorded, Rita's insouciance duly noted.

We are not conducting illicit insect races during these last summer evenings. No money has changed hands. Rather, we're trying to discover whether a group of *Forficula auricularia*, common earwigs chosen at random from our farm's seemingly endless supply, demonstrate traits that suggest each has an individual, identifiable personality.

There are two reasons for this mission: first, to cure our eleven-year-old of his only animal-related fear; and second, because personality, I've decided after a week's scientific reading has left me scrambling for clarity like a dozen startled earwigs, is an essential ingredient for having something to say.

First, though, I have a confession to make. Science scares me. Not in an exploding-Bunsen-burner or gruesome-medical-condition sort of way. Rather, through a quiet intimidation, a sense of not being clever enough. Science, I've always believed, is a place of truths and absolutes. It has right answers and wrong answers, and if you don't know the right ones, then you fail.

Once, when I was a little girl sitting at the dining table, I tried to join in with the adults discussing a politician.

"What an idiot," said my father. "Where does he come up with these ideas?"

I attempted a knowing laugh. "It's not as if it's by osmosis."

My father glared. "You probably don't even know what osmosis is," he said.

Chastened, I went back to my supper. He was right. I knew it had something to do with roots and shoots and nutrients, but I wasn't sure what. Science was for scientists, I thought, and I wasn't one.

It's with great trepidation, then, that I spend the first week of September outdoors, beneath china-blue skies, reading only science. Papers. Journals. Reports. Ethologists, biosemioticians, neurolinguisticians. I read and highlight, take notes, and read again. Because if I'm to discover what animals are saying to us, I need first to find out what *saying* really is.

Instead of answers, a plethora of questions emerge. What's the meaning of *talk*, of *say*, of *speak*? What constitutes language to begin with? What does *meaning* mean? I read articles titled things like "Do Animals Think about Thinking?" until I'm wrapped tight in knots, feeling confused, paralyzed, and as stupid as I feared. The last straw comes with a twelve-page paper on "the question of animal consciousness." By the end, I've lost my grasp of what consciousness is or whether I even possess it. Perhaps, I worry, I don't—or does this worry prove I do? Or am I just conscious of being unconscious, or vice versa, or neither, or both?

I default, unscientifically, to the dictionary. What, according to this more familiar pillar of certainty, is communication? What about talking, speaking, saying something?

Communication, communicates Oxford Languages, is "the imparting or exchanging of information or news…the successful conveying or sharing of ideas and feelings." To talk is to "speak in order to give information or express ideas or feelings; converse or communicate by spoken words." And to speak is to "say something in order to convey information, an opinion, or a feeling."

I switch, for variety, to *Merriam-Webster*. *Say*, it informs me, "1a: to express in words—state; 1b: to state as opinion or belief—declare; 2a: utter, pronounce; 2b: recite, repeat; 3a: indicate, show; 3b: to give expression to—communicate."

This clarifies things. Because setting aside the word *words* for later (since nothing, I'm to discover, says linguistic fistfight so much as the notion of animal "words" or "language"), I believe animals impart, convey, utter, pronounce, indicate, state, and communicate their opinions, ideas, and feelings to us humans all the time.

The piglets, meanwhile, do all of the above as their personalities emerge. Agnes, despite her initial Houdinism, is not an explorer; she likes to stay close. Irritated by affection if she doesn't instigate it, she's also more reserved at dinnertime, not, like her sister, doing her best impersonation of a short four-trottered kangaroo.

Chubby Constance, on the other hand, comes running when I call her and turns laps of joy like a puppy with the zoomies when let out to graze. A sucker for belly scratches, she'll keel over and lie helpless in paroxysms of joy. Constance purrs when she's happy; Agnes wags her tail.

Agnes learns immediately that her humans always emerge through the red side door. As soon as she hears it opening, she races to see, ears pricked and with an anticipatory snuffle. I try opening it silently to trick her, but her senses are too keen. Soon it's enough that she hears a footstep in the workshop on the other side of the door, then the sound of the inner door opening from the kitchen, then sounds from the kitchen window.

I wonder how long it took the famous Pavlov to condition his dogs to salivate. Surely it was more than forty-eight hours.

But Constance is the fiend for food. One morning, jumping up at me, she gets her front feet stuck in my wellies. I lurch forward, and her breakfast catapults itself all over her sister. She licks it off Agnes without complaint. She's an optimist, an extrovert, pushy and loud, where Agnes is quieter, more standoffish.

Later, as I sit reading a book chapter sent to me by its author, philosopher Hanoch Ben-Yami, titled "Logic and the Boundaries of Animal Mentality" (animals, claims Hanoch, do not dream, cannot hope, and can't understand negation or disjunctive syllogism*), Agnes approaches and nuzzles my palm. *Got treats in this one?* she's asking. *No?* She tries the other. Nothing. Off she trots to a tasty patch of dandelions and clover.

I quickly learn you don't herd a pig; you lead her. Having let them trit-trot themselves into the kitchen, I try to jostle them back out again and am met with great squeals of indignation. I revert to the grain-in-bucket method. "Come on, girls," I cry and dash off. The not-all-that-Light-Brigade charges after me. Pigs must choose to follow, it seems. They do not like to be pushed. I feel we have more in common every day.

The horses, too, find their feet: Major literally, as his abscesses drain away. I feed him soaked beet pulp and apples from our trees. His bite marks heal. He seems happier, relieved even. He greets me with his idiosyncratic whinny, a weird admixture of cowy moo and deflating tire, when I bring him treats. Tears prick my eyes the first time I see him flat out, dozing in the meadow; horses, like most prey animals, will only lie down when they're certain they are safe. Andvari, however, has taken a backward turn: Change weighs hard on this nervous pony. He reverts to how he was when he first arrived six months ago at my friend's place, an anxious wreck on permanent high alert. He flees when I move a hand too quickly. He jumps when Siegfried the pheasant performs his territorial bugle call from the field. I take it slowly, try to re-earn his trust with his favorite baby carrots, give him space to come to me in his own time, of his own accord.

The dogs settle instantly. They spend idle days wandering the

---

*   A disjunctive syllogism, also known as modus tollendo ponens or eliminating alternative hypotheses, is the classic logical understanding that "if either X or Y is true, and the answer is not X, then it's Y." For example, I have a baby carrot concealed in one of my two outstretched fists. If it's not in the left hand, then it must be in the right. Animals, claims Hanoch, a charming professor with whom I chat about it on Zoom, cannot understand this; Agnes might beg to differ.

farmhouse, tracking the sun from east to west, plopping themselves blissfully into puddles of it. No further fur flies. Outside we try to befriend Dolores the groundhog with plates of strawberries, raspberries, dandelions. Dolores spurns our advances, preferring her day lilies. A week later, she tires of us altogether and one night decamps pointedly to a stump on the other side of the fence.

A family of chipmunks, Indy and I discover, dwells beneath the lilac bush outside our front window. We sit on the doorstep as they come and go, hopping adorably through the undergrowth. White-tailed deer graze the apple orchard. We watch them through our peripheral vision so as not to scare them away: The placement of our eyes in our heads is a dead giveaway of our predator status. "Eyes on the front, I hunt," recites Indy. "Eyes on the side, I hide."

A tiny, tuft-eared red squirrel likes to chit-chit-chit at us from his cherry tree. We read up and discover his call, usually known as a rattle, is not the general "Get off my land" we assumed it to be, but rather a specific, individualized message, more along the lines, Indy says, of "This territory belongs to me, the Mighty Sven."

I fill a hummingbird feeder for our minuscule, bejeweled tyrant, who flies twelve-foot arcs to impress the ladies, swinging low like a hypnotist's pocket watch. When the feeder runs dry, he hovers at eye level at the kitchen window as if to remind me of my duty of care. But could he *really* be telling me this, like Dolly when she flings her empty water bowl around the kitchen?

I find nothing in science to confirm it, though a report from the University of Edinburgh says rufous hummingbirds, close relatives of our rubythroats, have an exceptional memory for the individual blooms they've visited. Might he think me, I wonder, a funny sort of flower? But if bees can remember faces, then why not birds? I read a slew of studies describing how wild crows, for example, recognize both human voices *and* faces. In my favorite, researchers donned a rubber caveman mask to trap, band, and release crows, and a Dick Cheney mask simply to saunter through their territory. Two years later, the crow community, its younger generation included, still scolded and dive-bombed the caveman, but left lucky Dick alone,[†] the kind of cultural transmission that reminds me of being warned to steer clear of a neighborhood house in which something nasty once happened. Clever corvids can even, it seems, differentiate between a familiar and an unfamiliar human language: A 2020 Japanese study found that large-billed crows captured in urban areas responded with greater interest to unfamiliar Dutch than to familiar Japanese. And I comb through plentiful anecdotal evidence of hummingbird-human interactions in which ordinary people repeatedly report hummingbirds seeking them out to refill, as one commenter called it, their "magic flower."

I replenish the feeder and relish this possibly unimaginary moment of connection.

Mostly, though, I wade through the history of scientific thought on animals' capacity for saying something. And I find that though scientists have broadly agreed you can only *say* something if first you can *think* and/or *feel* something, this is where agreement ends. Instead I discover a pendulum arc that swings back and forth, like our little hummer's, over whether and to what extent animals can talk, think, or even feel at all.

First, in humanity's earliest years, came Indigenous cultures who believed animals could talk and humans could listen, though not with human words or even necessarily with human ears, beliefs that occurred with compelling similarity wherever on the planet there were people. Dating back some forty thousand years, humankind's

---

†   Ethologist Konrad Lorenz, in *King Solomon's Ring*, describes a similar situation in which he took to banding young jackdaws while wearing a devil costume, complete with horns and forked tail, to prevent the birds from permanently fearing him. On his neighbors in his sleepy Austrian village, it appeared to have the opposite effect.

earliest known figurative cave paintings in Sulawesi, New Mexico, and France all include therianthropes: half-animal, half-human beings, which some scholars believe were efforts to express through art the connectedness early peoples felt to nonhuman species. To gaze at these startling, numinous images today is to experience a visceral thrill, imagining the depths of listening to which our ancestors might have been privy.

Next, swing forward a few millennia to the ancient Greeks and Aristotle, who, within their scala naturae (God and angels on top, plants and minerals at the bottom), believed humans possessed minds and the capacity for rational thought, but that animals did not. "The animals other than man," wrote Aristotle in his *Metaphysics*, "live by appearances and memories, and have but little of connected experience; but the human race lives also by art and reasonings."

The Middle Ages swung the pendulum the other way. Animals had both mental and moral agency, though the line was drawn at the possession of souls, and they were regularly prosecuted for the privilege—a pig for murder, a donkey for adultery—while in 1478, "pasture scarabs," chafer beetle larvae, were excommunicated by Benoit de Montferrand, Bishop of Lausanne, for damaging Swiss meadows.

Swing forward next to the Age of Enlightenment, when philosopher René Descartes tidied life up into mind and matter: the human brain on one pan of the scales and the entire rest of the world on the other. Animals, he said, were automatons who acted on impulse and instinct; you could even dissect them alive, because their cries of pain were purely mechanical. Only humans, Descartes reassured us, had thoughts and feelings and something to say.

"For it is highly deserving of remark," he wrote in 1637, "that there are no men so dull and stupid, not even idiots, as to be incapable of joining together different words, and thereby constructing a declaration by which to make their thoughts understood; and that on the other hand, there is no other animal, however perfect or happily circumstanced, which can do the like… And this proves not only that the brutes have less reason than man, but that they have none at all. For we see that very little is required to enable a person to speak."

"Cogito, ergo sum," Descartes famously concluded. "I think, therefore I am."

Animals didn't, and weren't.

Two centuries later, the pendulum arced back when naturalist Charles Darwin refuted all this. Evolution, he said, dictates that we're separated from animals not by some great schism, but by degree. "The lower animals," he summarized in *The Descent of Man*, "differ from man solely in his almost infinitely larger power of associating together the most diversified sounds and ideas."

Naturally, not everyone agreed. In the late nineteenth century, psychologist C. Lloyd Morgan came up with Morgan's Canon, a sort of animalic Occam's razor: We should never, he said, "interpret an action as the outcome of the exercise of a higher psychical faculty, if it can be interpreted as the outcome of the exercise of one which stands lower in the psychological scale." In other words, if behavior, including communication, can be put down to instinct rather than conscious decision, then instinct it is. Others objected to Morgan's objections: Your dog nudging at your back door, said Austrian ethologist Konrad Lorenz in his delightful 1949 *King Solomon's Ring*, is doing so purposefully and specifically to influence your actions.

Swing forward now to the 1950s and the behaviorists, led by psychologist B. F. Skinner, who harked back to seventeenth-century philosopher John Locke. Beings are born a tabula rasa, he said: a blank slate. Behavior, including language, is the product of external conditions. It can be *learned*. This was the era of putting birds in boxes. Skinner boxed pigeons, wherein they learned to peck a lever to receive a reward.‡ At Cambridge, zoologist W. H. Thorpe boxed baby chaf-finches. Reared in total isolation, they produced frail, feeble songs and, having missed their "learning window," could never produce the complex songs of a socialized bird.

Then came 1959, when a young student named Noam Chomsky wrote a devastating review of Skinner's latest book. No, no, he said. The capacity for language is innate, and genetic, and peculiar to man alone; a "universal grammar" is part of our unique human fabric. "It's about as likely that an ape will prove to have a language ability," Chomsky's famously quoted as saying, "as there is an island some-where with a species of flightless birds waiting for humans to teach them to fly."

Skinner, in the best academic tradition, ignored this "unknown character," then pooh-poohed him, then insulted him, but Chomsky refused to go away. Language, he persisted, is the result of a genetically determined, uniquely human program.

Things got nasty. Televised debates were proposed and impolitely declined. *Newsweek* published a letter from a reader who entreated it to "Locke up Skinner and give Chomsky Descartes blanche." A clear winner was never declared. Still, Skinner's ideas persisted. Some in the '60s tried teaching apes and dolphins and parrots in laboratories to use human language. Others put more birds—mynahs this time—in boxes. Biologists like Jane Goodall and Roger Payne, meanwhile, struck out into the wilds to study chimps' and whales' communica-tion in situ.

By the 1980s, the pendulum had settled on a sort of Darwinian

---

‡   Even more sinister than it might sound, this research formed part of Project Orcon, short for Organic Control, a pilot project designed to train pigeons to man (or pigeon) guided mis-siles. Though the pigeons performed impeccably in tests, the plan was never implemented. "Our problem," Skinner lamented to *Time* on September 20, 1971, "was no one would take us seriously."

gray scale on which every scientist picked their position. Linguistic purists continued to consider language an exclusively human domain. In response, some ethologists began using terms like *communication system* in place of *words* and *language* when investigating nonhuman species. They hadn't time for semantics, they said: It was what animals *could* do that was interesting, not what they couldn't.

Others still questioned whether animals possessed conscious thoughts at all, a phenomenon Harvard zoologist Donald Griffin, in his 1991 book *Animal Minds: Beyond Cognition to Consciousness*, called "mentophobia": an intense, deep-seated, and irrational aversion to the scientific study of nonhuman consciousness. Griffin and like-minded colleagues marked a shift from focusing on difference toward examining similarity: What have we humans in common, they wondered, with the world's manifold species? In 1998, affective neuroscientist Jaak Panksepp answered this question by identifying seven primary emotions: SEEKING/EXPECTANCY, RAGE/ANGER, FEAR/ANXIETY, LUST, CARE/NURTURING, PANIC, and PLAY/SOCIAL JOY.§ All mammals possess them, he said; they originate not in the cerebral cortex, the cognitive "human" part of the brain, but in the amygdala and hypothalamus, ancient brain structures we share with most animals.

In my experience, though, even *non*mammals demonstrate Panksepp's seven categories. My chickens, for instance, expected, sought, feared, nurtured, got angry, panicked, played, and experienced joy (we had no rooster in residence, but I expect the ladies might have lusted if we had). Ivy, our imperious Plymouth Rock, strode regally through our living room each morning on her way to the garden, having trained Agador, our 170-pound French mastiff, by way of several sharp looks and one sharper beak, to cower quivering in a corner.

---

§ Panksepp capitalized these labels in order to distinguish the seven primary emotional brain systems they describe from the use of these words in everyday language.

En route, she would strut, an avian Mae West, into the kitchen in expectation of her just deserts. "Oh, Amelia," I could almost hear her commanding. "Peel me a grape."

Naturally, not everyone would agree. Psychologist Herb Terrace at Columbia University, whom we'll meet in person later, maintains a Cartesian stance: Animals' utterances are instinctual (think: yelling in pain if someone steps on your toe), not intentional (creatively expressing your displeasure at whoever did the stepping). With no capacity for words, they will *never* have anything worthwhile to say.

Research and debate continue. Science continuously discovers there's more to be discovered. The year 2001 saw the discovery of the *FoxP2* gene, hailed as Chomsky's elusive human language gene until it was found in rodents and fish too. In 2014, neurobiologist Erich Jarvis and colleagues found that songbirds' neural circuits, which allow them to learn and produce novel songs, match those used for language learning in human genes.

And so we arrive at today, when if you ask a dozen scientists and linguists what feeling, thinking, and saying something is and whether animals do it, you'll get thirteen different answers. And therefore I come to some bad news and some good news.

The bad news: There's no one right answer.

The good news: *There's no one right answer.*

I'm astonished—but also instantly less scared of science.

All right, I decide. If no single scientific definition of animal "speech" exists, I might as well, as an unqualified person who's still not even sure about osmosis, go back to those simple dictionary definitions. Because my bar is not linguistic. I'm not interested in debating whether animals possess "true" language, because you don't, I believe, need spoken language to talk. Anyone who has ever bought a subway ticket in Tokyo without a lick of Japanese knows you don't need to share words to be understood. And if you've ever been locked out of the house, motioning with arms full of groceries to a teenager listening to loud music to let you in, or started a Zoom call with your microphone and headphones accidentally muted, you know plenty can be said without any words at all.

On my bookshelf sits a yellowing telegram dated April 3, 1942. CABLE AND WIRELESS LIMITED, it reads. At the bottom, a bottle-green map describes the myriad cable routes between Wellington and Dresden and Zanzibar and Talcahuano.

ALL WELL AND SAFE KEEP SMILING BEST WISHES AND GOOD HEALTH MERVYN TUGWELL, says the message. An inky postal frank shows it passed the army censor's inspection.

When my grandpa died in 2002, I found this telegram tucked between old diaries and daguerreotypes. I don't know who Mervyn Tugwell was. There is no one left to ask. But I've carried it around with me ever since, this brittle square that communicates so much, in so few words, about comfort and reassurance and war and worry and the things you can and cannot say.

Talking, to me, consists, like Mervyn's wartime telegram, of four things: a sender, a potential recipient, a means, and a message. The message doesn't have to be elaborate, nor is it, even when brief, necessarily straightforward. And the means isn't always linguistic: Anyone who's ever read *Paddington* knows the power of a hard stare.

We, the recipient, then take this message and interpret it subjectively, based on our knowledge of the sender and our mutual language (or lack thereof), along with our mood, personality, and a whole casserole of individual life experiences. We might be right in our interpretation; we might be approximate. Often we're entirely wrong.

But there's another requirement to have all this transformed meaningfully into saying something: That message must be sent from one *someone* to another. A machine, for instance, can both produce a message and has a means of communicating it: Ask Siri to come up with a joke or ChatGPT to compose you a limerick, and both will do so, then convey it to you via text or simulated speech. That, to me, is not truly saying something. For my purposes, the sender needs to be a living individual ("a single organism as distinguished from a group," says the old-faithful *Merriam-Webster*) in possession of a personality ("the totality of an individual's behavioral and emotional characteristics")[5]: that heady blend of traits, foibles, flaws, responses, feelings, and habits that makes them, and you, as unique and irreplicable as a snowflake or your fingerprint or a single grain of sand.

I don't doubt, despite the brevity of his communication, that Mervyn Tugwell was an individual with a personality. We all know our pet cats and dogs are too. The piglets, though raised in the same litter, have very different behavioral and emotional characteristics. But what about Siegfried the pheasant? Sven the cherry-tree squirrel? Our little imperial hummingbird (who, I decide, I'll call Mervyn)?

How small can personality go?

And that's where the earwigs and the dark room and the clipboard, along with a friendly Spaniard named Isaac Planas-Sitjà, come in.

The night we move into the farmhouse, which hasn't been inhabited in several months, there are earwigs everywhere. It's 10:00 p.m. and sultry, and historic rains all summer have brought insects out of the

---

[5]   Another useful way to think about personality is as *a consistent display of individual behavior in response to any given stimulus*. If this same behavior is broadly exhibited across the species, it might be a species behavior, not a personality trait. For example, if a human runs away from a bear, that's not necessarily because they have a fearful personality. But if they always run away from a butterfly, this could be considered a personality trait: You might label them "strange," or, if you have a kinder personality, lepidopterophobic.

woodwork. Outside, fireflies flash their come-hither bottoms. Inside, earwigs slither across the kitchen floor. They hasten along the baseboards. I swish open a living-room curtain, and a huddle of them, black, shiny, pincery, asquirm, falls out.

This presents Indy with a problem.

Because though Indy, whose heroes include Steve Irwin and Gerald Durrell, loves all creatures equally, he is, it transpires, terrifically afraid of the humble earwig.

"Why?" I ask.

He doesn't know.

I suspect it's the name. For despite entomologists' claims that earwigs are so named for their hindwings' resemblance, when unfolded, to the human ear, more common are the apocryphal tales of their crawling into those ears, getting lodged there, and laying eggs.

"They're the only kind of bug," Indy says in a small, quavery voice, "that I don't really like." He then climbs into a cupboard. "To read," he clarifies, and shuts the door behind him.

I think for a minute, then go for my *Descent of Man*.

"'Parental affection, or some feeling which replaces it,'" I read to him through the cupboard door, "'has been developed in certain animals extremely low in the scale, for example, in star-fishes and spiders. It is also occasionally present in a few members alone in a whole group of animals, as in the genus *Forficula*, or earwigs.'"

Silence. Then eventually: "So they're good mummies?"

"If you disturb one's nest," I add, "she'll even move her eggs."

Slowly Indy slides out of the cupboard. He watches an earwig skate across the pantry floor. "They're probably more scared of us," he concludes breezily, "than we are of them," and goes off to find his magnifying glass to take a closer look.

But are they? Could they *really* have emotions or personality, these multitudinous slithering beasts?

The following night, as cicadas crackle and a coyote howls to its pack, I call Isaac Planas-Sitjà, assistant professor at Tokyo Metropolitan University, to try to find out. A cheerful, goatee-bearded Catalan, he brims with enthusiasm over another much-maligned creature: the American cockroach, or *Periplaneta americana*.

They're dirty, people say. They bite. They spread diseases. Cockroaches, along with Twinkies, would be the only things to survive a nuclear holocaust.

But Isaac didn't decide to study cockroaches to dispel myth and rumor.

He decided to study cockroaches, he tells me, because of Sherlock Holmes.

"One night, while reading *The Sign of Four*," he explains, "I came across this phrase: 'While the individual man is an insoluble puzzle, in the aggregate he becomes a mathematical certainty.' That's exactly how we study insects, I thought: as aggregates and certainties. But I wanted to prove Holmes wrong. That even insects within a group are all *different*."

His most fascinating experiment, he says, offers a group of cockroaches, contained within an "arena," a choice of two shelters: one very dark, the other very light.

"They *should* all choose the dark one, because cockroaches don't like light. But they don't. Some individuals explore the arena more, even when most of the group goes for the dark shelter. I was expecting the collective to have this very powerful effect, as Holmes said—that you wouldn't be able to differentiate between them. But you still can."

In other words, explains Isaac, these cockroaches display the rudiments of personality. "That is, these differences within individuals are consistent over time and context. You can test them time and again, in different places, and you'll still see the same thing. In my cockroaches, this shows up primarily as bold versus shy, social versus nonsocial, risk-taking versus risk-averse. Social cockroaches tend to be shy and risk-averse, for example, whereas bold cockroaches tend to be risk-takers, but less social."

He doesn't yet know, he continues, why individual cockroaches are so different. "It could be genetic. It could be something they learn." I imagine Chomsky and Skinner, arms folded, at the center of a floodlit cockroach arena, preparing to battle it out. "Maybe there's a benefit to the group to be composed of different personality types."

"And what about emotions?" I ask. "Do cockroaches have those too?"

They exhibit, he answers, what some scientists call emotional primitives: basic, homeostatic emotions like fear or happiness, designed to keep them alive and far simpler than our often complex human ones, such as regret, or wistfulness, or guilty pleasure.

"When they start cleaning their antennae, you know they feel calm. When they run blindly into a wall, you know they feel scared. Sometimes we put them to sleep with CO_2, and I know they're stressed about it, and I feel bad. You see they're like, 'What's happening to me?'"

Isaac's work has given him a whole new appreciation for the American cockroach.

"One night, I was riding my bicycle home in the dark," he says, "and I saw a cockroach on the road. I crashed my bike to avoid it."

I'm hoping our less meticulous work will have the same effect, barring the bike crash, on Indy.

We run ten tests throughout the day with our dozen earwigs, whom we decide are all ladies. We use a big green recycling bin as our arena and a plastic plant pot with a door cut out of it as our shelter. The hardest part lies in telling the earwigs apart: We go by length, by the shape of antennae and cerci, by darkness or lightness of their six legs, and give each a name. We are the home guard of scientists: the ones wielding a margarine tub and a bendy purple ruler where everyone else has a beaker and sterilized calipers. Our notes are scribbled, our methods haphazard.

Nevertheless, we see a trend. Rita, we discern, is bolder. Mabel is shy. Gwendolyn prefers the shelter, whereas Cecilia tries repeatedly to climb the walls.

Indy is delighted, and so am I. With this simple act of something-like-science and a few Darwinian words, we've both discovered personality and overridden the power of human language.

All right, I think. What if we go *smaller*?

I put in another call, this time to Scott Waddell, professor of neurobiology at Oxford University, to find out.

Scott is a wry, friendly Scotsman, jet-lagged by his recent return from a conference of the Pavlovian Society. (I imagine a roomful of meringue enthusiasts and am disappointed to learn it's a gathering of animal behaviorists.) Nonetheless, he can summon excitement for his creature of choice: *Drosophila melanogaster*, the humble fruit fly.

Fruit flies in laboratories, he tells me, are bred to be genetically identical, "essentially clones of each other," to make testing as standardized as possible. Their behaviors should therefore be predictable: If one does something, they all should. Except between 20 and 40 percent of the time, he tells me, they don't. "You might find one who runs toward an odor when the others run away. And you'll find some flies who learn quickly and others who learn slowly. Some who, when given a choice in a T-shaped maze, will always go left and others who'll go right. And it's not genetically heritable: If you take all the lefties, it's not that you'll only get lefties among their progeny. You'll get a distribution again."

"So what is it?" I ask. "Why are they different?"

"It's because of 'noise' during brain formation, even in genetically identical creatures," Scott tells me, "that just slightly modifies the wiring: what a 2020 study describes poetically as the inherent chaos of normal development." This study gave fruit flies a straight line to follow. "A set of neurons in the visual system," it found, "is wired up in a variable manner, resulting in brain circuit asymmetry unique to each fly who guides its line-walking behavior." Asymmetrically wired flies walked straighter; symmetrically wired flies zigzagged.**

I linger over this invigorating idea—that from fruit flies to piglets to you and me, this unpredictable, uncontrollable dice roll, before Chomsky's nature or Skinner's nurture, governs who we all are.

"But that's not all," Scott says. "Fruit flies, like cockroaches, also possess those emotional primitives. I believe there are mechanisms in the brain that are the foundation of emotions, but it's just that with animals, we can't tell if they have subjective feelings because they can't express them to you."

Take facial expressions, he says.

"They're obvious in mammals, but aren't something flies are good at. Yet they'll withdraw their proboscis when things taste bitter and extend it when things taste sweet, just as you'll see pictures in scientific literature of human babies scrunching up their faces when something tastes bad."

Though fruit flies appear to be unlike us in every conceivable way, they are, he says, "more like us than we'd ever imagine. It seems the mechanisms our nervous systems use are similar. Dopamine, the feel-good chemical, for instance, is very heavily involved in the learning experience of every animal, humans and flies included."

And fruit flies, he says, like bees and humans and hummingbirds, form memories too.

"My lab devised ways to see memories form in the fly brain. We can even implant memories into them, to have them behave according to experiences they've actually never had. We can give them reward memories by activating specific neurons, for example, even though they've never experienced that reward for real."

** In another report to this effect, the mating choices of male earwigs, who each have two penises, were studied in 2021 in Japan. To researchers' surprise, earwigs randomly favor the use of one of their pair of penises, being either a "southpaw" or a "righty." To study this further, scientists amputated just one of each earwig's penises, only to find he was just as fecund with the remaining, previously lesser-used one. I decide not to tell Indy just yet about this fascinating bit of weird science.

"Like *Total Recall*?" I frown. "For flies?"

He grins. "In the twenty-odd years I've been in the field, it's all become quite sci-fi."

Still, he says, there are an awful lot of researchers who are reluctant to believe that insects are not just reactionary devices.

"That goes right back to the dawn of philosophy and defining other animals as beasts to justify eating them, because they aren't as sentient as ourselves. And I think that's really been maintained in people's views of insects. But it never ceases to amaze me what they're capable of. I think everything we've tested them for, they've pretty much been able to do it. And often we've found they've done something smarter than we had imagined."

This reminds me, I tell him, of that study I read on fruit-fly dialects. "We find flies can communicate with one another about an anticipated danger," it said, "which is suggestive of a fly language."

"Exactly," Scott says. "Mad stuff."

Later, in the garden, our crack team of scientists releases the earwigs. They head off in different directions. Mabel makes for the house. Gwendolyn goes up a wall. Rita doesn't seem too motivated to head off at all.

Yes, Indy and I agree, these *Forficula* are all individuals, possessing personalities. But even if we found they had the means, could they be sufficiently conscious of our existence to ever have a message specifically for us? For a start, says Indy, surely the size discrepancy is just too vast. It would be like a human having a casual chat with Kanchenjunga.

As Mervyn zips from feeder to raspberry patch and back again, we sit drinking lemonade and together come up with two additional Saying Something criteria, both concerning this tricky territory of consciousness.

Number 1, we decide, the individual with the personality must be conscious that you, the recipient, actually exist (and that they, the sender, exist too).

And number 2, adds Indy, their *message itself* must be conscious: You've got to intend to be doing it. It's not talking, he elaborates, if you don't realize you're saying something, and we humans send unconscious messages through body language all the time.

"Like what?" I ask.

"Well, I've read that if you look up to the right when you're talking, you're probably lying."

And there's also mirroring, I add. That state of limbic synchrony where we subconsciously mirror a conversation partner's posture and gestures to make them feel comfortable or understood.

That deals neatly, we decide, with criterion number 2. But what about number 1? How can one tell? And what exactly *is* consciousness anyway?

As I go for Google ("Consciousness: the state of being awake and aware of one's surroundings…the fact of awareness by the mind of itself and the world"), Indy decides he's got better things to do with his Saturday and leaves me to it. I take Puff alone to the orchard to sit and ponder further.

The day is overcast. Siegfried trumpets from the field. The carpet of dandelions underfoot, I notice, have not opened their glorious golden faces, though they didn't consciously choose not to. Philosopher Thomas Nagel, in his seminal 1975 essay "What Is It Like to Be a Bat?," wrote that "An organism has conscious mental states if and only if there is something that it is like to be that organism—something it is like for the organism. We may call this the subjective character of experience."

For a rock or a dandelion, I ponder, possibly not so much. But for Puff or Siegfried or you or me, of course: We're all conscious we're our own individual beings, moving through a world of other individuals we know to be separate from ourselves.

Neuroscientists agree. The 2012 Cambridge Declaration on Consciousness, signed by participants of the Francis Crick Memorial Conference at Cambridge University, states that "humans are not unique in possessing the neurological substrates that generate consciousness. Nonhuman animals, including all mammals and birds, and many other creatures, including octopuses, also possess these neural substrates." The 2022 British Animal Welfare (Sentience) Act reiterates this, recognizing animal sentience, or consciousness, in law for the first time.

The orchard feels peaceful. I sigh and glance at Puff. She glances back, blinks, and without warning launches herself sideways at me to try to wrest the scrunchie out of my ponytailed hair.

"Hey! Cut it out!"

Instantly, she stops. She wags her tail and licks my hand. *Just a joke,* she seems to be saying. In this moment, I recognize, Puff is intentionally communicating. She is an individual with emotions and a personality, a recipient, a message, and a means. She's conscious she's her, and that I am me, and that I have something she wants that I don't want her to take. She satisfies all eight of our Saying Something criteria.

But there's still one last, tantalizing test of consciousness I've read about that I can't help but want to try out.

Time for another stab at science.

The cusp of fall brings stormy days to the valley. Major, who hates getting wet, dozes beneath his shelter. Winnie watches contemptuously from the window. Only Andvari stands out in it: a Viking in all but valor, he loves inclement weather.

I, meanwhile, am indoors in the workshop, putting lipstick on a pig.

Outside the farmhouse, thunder rumbles. The downpour, the radio reports, has washed away roads and driveways across our county. What better moment to test for cognitive self-recognition, the awareness by the mind of itself and the world, in my otherwise unoccupied animals?

In 1978, psychologist David Premack produced a paper titled "Does the Chimpanzee Have a Theory of Mind?" "An individual," he wrote, "has theory of mind if he imputes mental states to himself and others." Premack's paper makes for lively reading. It's as if I were at a cocktail party in a sunken living room, gimlet in hand, listening to Premack, in polyester, explain his groundbreaking new study.

His premise was to play chimps films of human actors struggling through practical problems: some simple, like trying to grab an out-of-reach banana, others complicated, such as being "unable to play a phonograph because it was unplugged." He then showed them a series of photos, only one of which was the solution to the quandary. The chimps invariably picked the right one (except when a chimp named Sarah, wishing one particular actor ill, chose the photo of him "strewn with cement blocks"). This, said Premack, proved chimps possess theory of mind: "The chimpanzees' consistent choice of the correct photographs can be understood by assuming that the animal recognized the videotape as representing a problem, understood the actor's purpose, and chose alternatives compatible with that purpose."

But do other animals have it?

Premack tested whether chimps could recognize the motivations of an individual who was *not* themselves. But another way to look for the presence of consciousness is to investigate whether an animal possesses a concept of self in the first place. In 1970, Albany psychologist Gordon Gallup developed the mirror test to explore just this. He would, he decided, mark animals' faces with dye and see whether they'd investigate this anomaly in the mirror. Like Premack's tests, it worked perfectly on chimps: "Insofar as self-recognition of one's mirror image implies a concept of self," Gallup concluded, "these data would seem to qualify as the first experimental demonstration of a self-concept in a subhuman form."

Thereafter, only orangutans and humans, he found, passed muster. Fifty years later, and though other experimenters have concluded otherwise, he still sticks by his claim: "Based on rigorous, reproducible experimental evidence," Gallup wrote in 2020, "only some great apes and humans have shown clear, consistent and convincing evidence that they are capable of correctly deciphering mirrored information about themselves."

I decide to give it a whirl anyway.

I take down the mirror from above the fireplace and prop it against the workshop wall. Immediately, Puff trots over to see what I'm up to.

Before I can even dab on the lipstick, she freezes at the sight of her reflection. She cocks her head, goes to bark, stops, and looks again. She takes one step to the right and observes her reflection do the same. She stands stock-still for almost a minute, just staring, then sits down, her baggy bully body soft but alert. She looks at me in the mirror, then at me in real life, as though she were trying to understand that *this* is *her* and *that* is *me*, but so are *they*. Occasionally, she glances from my mirror-me's eyes to the back of my actual-me head, to her own mirror eyes and back again, as if confirming her suspicions.

Then she yawns, shakes, and wiggles over to be petted.

Dogs have historically "failed" Gallup's test, and I don't know what Puff was really thinking. Still, I remember doing something similar as a child: looking deep into the mirror until the moment things started feeling very strange, as if my mind were struggling to fathom the unfathomable. (I challenge you, even as an adult, to wrangle with what's known as the Caputo effect, as our brains wrestle with the question "Who the hell is this anyway?") I wonder if it's a primitive reaction, harking back to a misty prehistoric past when we first glimpsed our reflections in some glacial body of water and recoiled at the sight.

I tempt the piglets out from their pigloo to try it on them, but they're more interested in a football they find in a corner than in the sight of their plumptuous selves. I apply scarlet bindis to each forehead and lure them back to the mirror. Though pigs, according to a 2009 Cambridge University study, are "highly sensitive animals who become aware of their own existence," their eyesight is poor: Constance has twice mistaken Indy's fingers for baby carrots. I'm not surprised they are unintrigued by their beautified reflections.

So I try Dolly. I try Winnie. I lug the mirror out into the rain to test the horses. But aside from one exploratory lick from Andvari, who seems disappointed it tastes only of mirror, nobody cares about their reflected selves, with or without a lipstick beauty spot.

Nevertheless, as astronomer Carl Sagan said, absence of evidence is not evidence of absence. And some scientists feel it's the science at fault, not the subjects. A paper I come across titled "Cultural Variations in Children's Mirror Self-Recognition" suggests many small children in non-Western countries, including Fiji and Kenya, "fail" the mirror test too. More anecdotally, in 2018, a Scottish fold kitten named Mimo caused a stir online when, in a short video clip, she balanced on her hind legs to stare into the mirror, seemingly watching herself groom her own ears.

Other scientists have succeeded with species-specific experiments. At the New York Aquarium, psychologist Diana Reiss's team marked bottlenose dolphins on parts of their bodies they could feel, but not see, and watched as they swam to the mirror to take a look. Dr. Alexandra Horowitz, with whom I'll chat later, performed a successful olfactory mirror test for dogs wherein they sniffed out their own pee differently from that of their conspecifics. This version, I decide, I will not be attempting.

And then there are voles, like the sweet, scurrying family who live in tunnels beneath our front doorstep. Voles, says a 2016 study, demonstrate empathy: An unstressed vole will exhibit "consolation behaviors," spending more time than usual grooming its stressed mate. That seems a lot like theory of mind to me. Likewise, jays and crows, who choose particular foods or gifts they believe will appeal to prospective partners or to benevolent humans who feed them. One could argue they're just guessing, but that only makes them like every husband ever, and superior to my university roommate, who bought his unimpressed mum an ironing board for two successive Christmases.

Perhaps, I conclude, we cannot hold fast to a series of tests to decide whether an individual is capable of meaningful communication. How animals conceptualize themselves and one another, and how they show it, surely depends on how they perceive the world and their place within it. Little wonder, I think ruefully, that humans see self-awareness in their own mirror image, versus "subhumans," who do not.

I'm reminded of evolutionary biologist James Gould, who in 1975 proved to a skeptical scientific community the existence of honeybees' waggle dances. "Some of the resistance to the idea that honey bees possess a symbolic language," he wrote, "seems to have arisen from the conviction that 'lower' animals, and insects in particular, are too small and phylogenetically remote to be capable of 'complex' behavior... Despite expectations, however, animals continue to be more complex than had been thought, or than experimenters may have been prepared to discover."

Science, I'm learning, is like a game of hide-and-seek; just because it hasn't found something yet doesn't mean it won't. It was, after all, barely eighty years ago that Donald Griffin persuaded naysaying scientists that bats navigated via echolocation, something we all now take for granted, but so new and shocking a concept in 1944 that Griffin coined the very word. If fruit flies can possess personalities and dialects, isn't it just possible that little creatures could be conscious of themselves and of us, with some sort of theory of mind too? Maybe we just haven't done the right test yet; maybe theirs just doesn't *look* the same. And if so, we may still one day be asking how to improve our communication with a cockroach, just as we might currently wonder how we can better converse with our cockapoo.

I am sitting outside on an unseasonably warm day late in September, rereading Nagel's "What Is It Like to Be a Bat?" Honeybees buzz in the oregano. Indy lies reading *Peanuts* on the lawn. I watch two young ospreys above the river twine and tumble. *They know they are themselves*, I consider. *Separate in space.* Cabbage-white butterflies flit through the meadow. Moths and butterflies, I read recently, retain memories from before metamorphosis. Are these whites, incredibly, remembering life as caterpillars?

At the kitchen window, Mervyn prepares for his international departure, spending more time than usual at the feeder. The earwigs, like Dolores, have moved out: It seems we've communicated to them simply by our presence that indoors now belongs to us, and they have inadvertently communicated to us that they'll go elsewhere—even Rita.

My children's favorite picture book when they were small was John Burningham's *Would You Rather?* Would you rather, he asks, an eagle stole your dinner or a hippo slept in your bed? Would you rather live with a fish in a bowl or a rabbit in a hutch?

"I'd be a goldfish," Zeyah invariably answered.

"Why?" demanded Tyger.

"Because I'd get to swim around all day. I love swimming."

"Well, that's not *being* a goldfish. That's you being *like* a goldfish."

This conversation among the under-tens forms the crux of Nagel's essay—that even if animals have personality and possess intention, emotion, self-awareness, and a message they wish to transmit to us, there's no way we will ever understand them because *we are not a bat.* The only way to know what it's like to be a bat, Nagel argues, is to be a bat and have bat concepts.

The philosopher Ludwig Wittgenstein said something similar: "If a lion could talk, we could not understand him," because the lion's sensory and subjective perception of his world—his umwelt, in other words, a term popularized by German biologist Jakob von Uexküll —is so different from our own.

But is umwelt, I wonder, everything? We don't know exactly how even our closest friends or lifetime partners see kingfisher blue (which Cairo, our third child, maintains is green), or taste cilantro ("Soapy!" spits Cairo; "Herby," declares Zeyah), or experience love. Does your soapiness have to be the same as mine for us to find connection? My favorite conversations have almost always been with people who, ostensibly at least, are nothing like me.

I snap Nagel shut. I don't expect a conversation with an animal to be about human-umwelt things like poetry or philosophy. But that's not to say there's no value in it (conversing with my teenagers involves very little of those either). Indeed, maybe there's more to be gained from seeing things differently: Perhaps that's why nature intervenes, even when humans clone fruit flies, to make each one dissimilar. Instead of seeing Nagel and Wittgenstein as interspecies party poopers, I decide to see them as conversation openers. Language, after all, as I remember once reading in a *New York Times* comments section, is in the ear of the beholder.

Everything I've encountered thus far this month has convinced me that even the tiniest beings have something to say. But if we humans consider ourselves the best at talking, at thinking, at feeling, even at *being conscious,* and simply apply our own criteria to everyone else, then we're all doomed to fail an exercise in futility, trapped in our own stuffy echo chamber. I now know what I believe constitutes saying something: Though true wisdom—Plato said Socrates said, and Thoreau said Confucius said—is knowing you know nothing.

"If earwigs tested us by earwig standards, Mummy," Indy looks up from his comic as one slithers over Snoopy, "they might think we're really stupid."

I go into the kitchen for a cup of tea, where fruit flies patrol, no matter how hard we try to keep them out. Little amber bodies, flitting around, land on the bread, the butter, the bowl of bananas. Last year, I vacuumed them up without compunction. But now... I stand staring at one on the lid of the compost bin. *Who are you?* I find myself wondering. *Are you a bold one, or are you shy? Are you a lefty or a righty? Would you feel frightened as you struggled against the vacuum's slipstream? Would you communicate that impending danger to others before it was too late? Would they understand your accent?* I know I can never be a fly. I'm uncertain we will ever have a conversation. But I also know we are sharing this space. That we both experience fear and have memories and get dopamine rushes. That we're both individuals with dialects and personality.

Instead of the vacuum cleaner, I fetch Indy's magnifying glass. I examine the red eyes of this *Drosophila*, a cross between pilot goggles and licorice allsorts. Its whiskery, old man's back. Its intricate stained-glass wings. And it makes me think of Blake:

> Little fly
> Thy summer's play,
> My thoughtless hand
> Has brushed away.
>
> Am not I
> A fly like thee?
> Or art not thou
> A man like me?

Already this quest is changing me. Over the last three weeks, I have noticed more and more both the similarities *among* species and the differences *within* them. I've observed one orb weaver spider beside the front door, who always stays put when I approach, and a second, who reliably retreats to the far edge of her web. I've witnessed the vagaries of seemingly identical houseflies trapped indoors: some flinging themselves persistently at windows, others turning circles, apparently content, beneath the slow-moving fan. Outside, I watch an enormous black field cricket, a *Gryllus campestris* far braver than all the others, with his palpi stuck into the lump of breakfast burrito I accidentally dropped on the ground. He's tucking in, like I do, with gusto. He is no Cartesian robot. Does he know I am here? Would he tell me, "This is tasty," if he could? I cannot rule it out.

My own pendulum, I realize, has swung.

Early morning, mid-June 2020, and I'm feeding the horses at my boarding-barn breakfast, mucking out their stalls. Across the globe, the COVID-19 pandemic rages. Apart from my daily trips here, I go nowhere. Except for my husband and kids, I see no one. Borders are locked down, and I, like millions across the world, am missing my family overseas.

BZZZZZZ. My phone vibrates in my pocket. A text message. I already know what it's going to say.

My grandma in England, my staunchest supporter, with her scalding wit and three-inch stilettos, has been ill for several weeks. Over the last few days, she's refused food and, since yesterday, water.

I listen to equine mouths munching quietly before pulling out my phone.

Sis—can you call me? says the text from my sister.

I bite my lip and text back. Not much signal. Is it Grandma?

This isn't true: I just can't bear to talk, and it seems easier to communicate, like Mervyn Tugwell, in just a few characters.

Yes. I'm afraid it's not good news.

You can tell me.

She died peacefully this morning. I'm sorry, sis. Talk later?

I nod, tuck my phone away, and recommence shoveling shit.

It's a funny feeling when someone dies. Your head fights your heart, one telling you *They're gone*, the other *I'll talk to them soon*. You can't—won't—believe you'll never hear their voice again. Or will you? In your mind, you speak to them and hear a reply, but you also know you don't.

I set down my shovel and lean against the barn wall. I can't cry. I try to breathe, but a sticky ball of grief is stuck in my throat. I think of all the others going through this at this moment. *You're not unique*, I tell myself.

But still I lean here, my hair hiding my face, grateful for social distancing that will keep everyone away.

Then I feel something warm on my chest.

Major, I find, has left his breakfast and walked over to me. Now here he stands, his nose an inch from my sternum. I laugh, a strange, strangled hiccup. He never *ever* leaves his food, this horse of mine.

Gently, he drops his head until his forehead rests against mine. I feel his scratchy forelock, the solid surface of his skull.

And here we stand for half an hour, maybe more. Forehead to forehead. Breathing together. Silent, but not silent at all as grief and gratitude commingle, and I feel, beyond the fragmentation of loss, the connectedness of it all.

Does he understand the intricacies of human grief, this big red horse? I don't think so. But does he recognize, in me in this moment, an emotion he can understand and to which he can respond? I think he does.

Does he say so in words? Of course not.

Do we have a conversation? I think we do.

2

Fowl Language

Why Listen in the First Place?

> "So they show their relations to me and I accept
> them,
> They bring me tokens of myself, they evince them
> plainly in their possession."
>
> —WALT WHITMAN, "Song of Myself"

OCTOBER

On October 4, St. Francis's Day, I wake at 3:14 a.m. from a dream about talking piglets and type a groggy note into my phone. *Y*, it reads, *do i want 2 kitten 2 what animals r saying? learning love nature (Pippin n Pooch) curiosity (why say what say) safety (not get stomped ect) compassion empathy (wild animals farm animals all Annie mules).*

I doze until the first exploratory notes of birdsong float through the bedroom window. Then I rise and pad downstairs, past children (sleeping) and dogs (sleeping), glance out at piglets (sleeping), pour coffee, pull on wellies, grab my phone, and head outside.

Dawn slips in, coral and marigold. The starlings are black paper cutouts in the oak. Sea mist rolls upriver. Major greets me with a whoopie-cushion nicker as I bring the horses hay, then continue down into the meadow. There I settle on a mossy rock with my Merlin app and steaming coffee to prepare for a conversation later this morning with renowned soundscape ecologist Bernie Krause.

We humans, I consider as I take a first sublime sip, along with manifold other species say things in many ways, for many reasons. We convey information or instructions. We ask questions. We express, in varying degrees of complexity, our experiences of the world: of love, fear, desire, frustration, trepidation. But at our innermost core, every single one of us, from human to starling to whoopie-cushion horse, talks because we want to be understood.

In which case... I try to make sense of my witching-hour note... *Y kitten?*

Every day, twice a day, for almost a decade of my childhood, I took a walk with my next-door neighbor, Miss Mander, and Pippin and Pooch, her corgis. Wherever I was in our house—in my bedroom, in the bath—I would hear the dogs bark and come running, to range for hours with this spinster along the Staffordshire canals. Miss Mander didn't care what society thought. She had home-cut hair and wore kilts and turtlenecks. Together we tromped hundreds of miles across fields and down towpaths. We listened to birds and frogs and foxes. We witnessed the fantastical annual progression of spawn to tadpole to toad.

Listening to the natural world gave me a feeling of peace. It provided me with continuity and a sense of place, not only in space, but also in time: I could close my eyes and know what season, month, time of day, and sometimes even day of the week it was just from the sounds I could and couldn't hear. No matter how mean Mrs. Fox or Mrs. Badger were at school (the genuine articles, I decided, were both far superior), I found nature was always there, nonjudgmental, ripe for the listening, reminding you there's a world much bigger than yourself, your single perspective and experience. The only thing I didn't like was when I had to kiss Miss Mander's whiskery cheek in thanks for my annual birthday present. I still have one: a copy of *Alice's Adventures in Wonderland* with the original Tenniel illustrations of Alice listening to the Dodo, the Dormouse, the Cheshire Cat.

But like Alice, I grew up. I stopped running for the door when I heard the dogs bark. I started caring about friends and boys and parties. Sometimes I waved from the car window as I passed Miss Mander out walking her dogs alone. Pooch died, then Pippin. Gradually, a little connective window of listening creaked shut. The world became primarily human: physically bigger, but smaller somehow too.

We stop listening to the natural world, it seems, when we start caring more what the human one thinks of us. Now here I am, in a field at daybreak, in sensible clothes of which Miss Mander would approve, attempting to pry it back open.

Listen, I instruct myself. *What can you hear?*

First: birds.

I hear the gleeful music of song sparrows. A piping northern cardinal. Red-winged blackbirds. The starlings. A couple of crows. The hollow *clonk-clonk* of a hairy woodpecker.

My app hears things I don't, its spectrogram representation a cityscape of Gothic spires. It registers redstarts. Four kinds of warblers and another woodpecker, this one pileated.

My turn. I hear a blue jay. A goldfinch.

It contributes a yellowthroat.

I focus on other noises. The low of a cow searching for someone (a friend? a baby?) in the fog. The morning's first crickets, still *cree-cree-creak*ing as if trying to sing back summer. A distant tractor. A squeak of Sven, our red squirrel. The anticipatory *Wheeeee* of the piglets, hoping for an early breakfast. Myself, taking a noisy slurp of coffee.

A fly *pzzzes* past. I hear the pigeons on the barn roof, their claws tip-tapping along its metal spine. A shotgun report from across the river, causing Siegfried to shriek in the wood. (He's right to be horrified: Fifty million pheasants are shot annually, and his imperial garb, like a royal poorly disguised among the peasantry, doesn't exactly help.) And the occasional pop and slurp, not of coffee this time, but of mud in the meadow's ruts and hoofprints as it soaks up the morning dew.

A picture, textured and three-dimensional, composed of what Bernie calls the *geophony* (nonbiological natural sounds, like the wind or the river), the *anthrophony* (man-made noise, slurps of coffee included), and the *biophony* (the sound of all other living organisms).

"That is neat! That's really neat!" he exclaims when I tell him later about the sound of the dew in the hoofprints. "Did you record it? You must capture that sound. That is really cool. Just with your cell phone. Record it, record it!"

A spry eighty-five-year-old with a thick head of wavy gray hair, Bernie has spent the last forty-five years recording the planet's brimming-with-life soundscapes. It's vital to listen to the natural world, he says, because sound reveals clues to the health of an ecosystem that visual observation alone cannot.

Bernie grew up in suburbia, on the edge of 1940s Detroit before it exploded with industry. His mother mistrusted the natural world; it was wild, dirty, dangerous. But his earliest memory as a child "with poor eyesight and an acute case of ADHD" is of soothing birdsong outside his bedroom window.

He became a musician, composer, and sound engineer. He moved to Los Angeles, played the Moog synthesizer on a Monkees record, and, he tells me casually, worked for Coppola on the soundtrack of *Apocalypse Now*.

"It was ego gratifying and all that shit," he says. "But the real work was back in nature."

So he left town for a PhD in bioacoustics.

"At that time, the model at places like Cornell or the British Library of Wildlife Sounds was to record individual species out of context with parabolic dishes. I was coming to it as a trained musician, so the sound to me was only significant if it explained something about how it was being expressed cohesively, with all the animals vocalizing together."

Animals, he tells me, don't experience the voices of the world as a series of separate solos. "Nor did our ancestors, who could ascertain the health of a place just by listening. What were we missing by only listening to parts of the whole? I decided to put Humpty Dumpty back together again."

For decades, Bernie recorded the biophony at Lincoln Meadow, in the Californian Sierra Nevada. He shows me spectrograms from 1988, the year before a lumber company "selectively logged" the meadow, and from 1989, a few months after. The first spectrogram, like my own meadow's, is dense with activity: A poltergeist readout in a scary movie. The second is a flat line except, halfway through, for the pacemaker tap of a Williamson's sapsucker. To the eye, however, the landscape looked the same.

"We're fucking things up," Bernie tells me. "It's an environmental catastrophe. And sometimes you can listen to extinction much more easily than you can see it."

Back in the '60s, he says, it would take about ten hours of recording to capture an hour of natural soundscape for an album, art piece, or movie soundtrack.

"And now?" I'm expecting him to say that technology has reduced that time significantly.

"One thousand hours or more. There's human noise. There are fewer wild places. There's less life remaining in many of them."

I'm gobsmacked.

"Why listen?" he continues. "To beware the anthrophony: the insidious influence our own sound has on the environment. Like car noise from highways that prevents birds from finding each other with their calls. When synchronicity is broken like this, lone voices become vulnerable to predators."

I think of little boy Bernie in his Detroit bedroom listening to the perilous natural world outside when really, we're the danger.

"How can we listen better?" I ask.

"Go outside three times a day," he prescribes. "First, half an hour before dawn. Then late afternoon, when human-created noise tends to die down. Settle in and just listen. Finally do the same, in the same place, at night. Night," he adds, "is magical. It's how we got our spirituality. People didn't know what they were hearing, so they made up tales about gods in the forest."

"So in a sense, listening to animals created culture?"

He nods and smiles.

"Wow" is all I can say. I open my mouth. Close it. Open it again. "Wow."

Rachel Mundy, a professor of musicology at Rutgers University, knows all about the anthrophony. She lives and teaches in Newark, New Jersey, which some, she says when we talk the next day, would call "the armpit of urban America." "Don't quote me on that." She sports a big grin and overalls. "Find something more diplomatic."

All right. Newark, New Jersey, is an intensely urban environment. New York, but not *the* New York. Best known for its massive Liberty International Airport built on "reclaimed" marshland, it's not somewhere you'd naturally go to listen to wild animals.

Yet each spring, Rachel teaches a university class on precisely this kind of listening, in which, in just forty minutes, students learn to tell the difference between one American robin, that common brown bird with the kind, yellow-rimmed eye, and another.

"Each individual robin," she gives me the CliffsNotes version, "produces very sweet, melodic phrases, sort of 'Do-oo-woops' that happen very quickly. Within its lifetime, a bird will learn about twenty-five of these separate phrases. They're not all unique to that bird, but the set—the way they combine them—is. A bird will have a favorite phrase that it uses a lot. You'll start to learn, 'Oh, this bird really likes this phrase,' and maybe that phrase is an upside-down version of one of its neighbor's. They all sound *robin-y*, but they're still unique."

"Once you concentrate on listening to one bird at a time," she says, "you'll start to hear the pattern. At first, you can use a spectrogram or record it and play it back slower to help train your ear. But," she assures me, "it's like riding a bicycle. Once you do it once, you can do it anytime, anywhere."

The reason this is so special, she says, is that in urban settings like Newark, it's easy to move through the world as if the only species that exist are humans and pests. "Which is an us-against-them sensation. And I feel birdsong is an especially good gateway to start rethinking that. After students take that class, they never *don't* know there's a mockingbird on this one corner above a tire shop or a robin there every morning in front of the grocery store. And it's not 'robin the species.' It's an individual bird."

Unlike Bernie, who warns against the sound of silence, Rachel worries about the voices that surround us daily, which we fail to hear. "To me, the fear is not that you suddenly notice when there's nothing there. It's that you don't notice when something is. That's what I feel almost panicked about—the fact that so many of us, including me, have walked through our lives surrounded by other beings, but with a total lack of awareness of them. The most profound part of listening is recognizing that you're not alone: Whether you want it or not, you are never alone."

This notion rings a bell. After our call, I pull out Maurice Maeterlinck's lovely 1901 *The Life of the Bee*. "What have we to do," he writes, "with the intelligence of bees… The discovery of a sign of true intellect outside ourselves procures us something of the emotion Robinson Crusoe felt when he saw the imprint of a human foot on the sandy beach of his island. We seem less solitary than we had believed."

Then outside I go to listen again to our farm's own birdlife.

There are the starlings, now whistling their loony matins from the oaks. The *crah-crah*ing crows who've descended on the field to fling their daily dose of horse poo. The chickadees passing through, from wood to wood, with their gleeful *chickadee-dee-dees*. The rat-a-tat woodpecker in his black-and-white soccer strip. And our pigeon pair, *Columba livia*, the bonny country cousins of the dirty birds I liked to save as a child, with plumage so bright that Gal first thought them parrots, and Indy, on the dark side of dusk, a pair of shooting stars.

I watch them flap down into the horse corral, where they strut about, tossing hay.

Some ornithologists think pigeons first arrived in North America from France in 1606 at Port Royal, Nova Scotia, only a mile downriver from me. I wonder if this couple are direct descendants of those four-hundred-year-old forebears.

Brrrrrrr-oooo.

Brrrrr-oooo.

I listen, trying to tell one from the other. One tenor, one contralto, but who is who? Or are they alternating? I find it impossible to say. But while listening, I notice they look as different as Constance and Agnes. The female is sable, with terra-cotta feet and white wings connected by a band across her back, like an angel in a school play. The male's feet are the color of Fanta. His throat is oil-slick green, with a white stripe above his beak and a coal-dust body. How did they choose each other, I wonder, these two who have paired up for life? They're not pigeon the species, I think. They are pigeondividuals.

From the edge of the wood, a crow call erupts. *Rah! Rah!* Instantly, our couple takes flight. They sweep a graceful parabola, pinks flashing, back up to the barn. I look about and see no cause for alarm. But twenty seconds later, a bald eagle glides over. The pigeons, I notice, keep completely quiet. The eagle flies on. A minute later, they're back down, and I have discovered something new about them: their ability to understand, and benefit from, the communications of another species.

For much of *Homo sapiens* history, we have sought out the company of animals. Thirty thousand years ago, we began domesticating dogs; ten thousand years ago, sheep, cattle, and pigs; and six thousand years ago, horses. (I might, though, dispute the claim that eight to twelve thousand years ago, we domesticated cats and suggest we still haven't fully succeeded.)

We aren't unique in our penchant for cross-species relationships. There are symbiotic relationships between cowbirds and cattle, between crabs and anemones. The Dutch airline KLM, I was told long ago by a Dutchman, once had chickens on its payroll as companions to air-traveling racehorses; my own less high-flying red horse lived with a miniature donkey to whom he grew so attached that I'm still considering buying him one of his own. Milo, our schnauzerish, and Charlie, our wallaby, learned to play together, though their play styles (Charlie favored grabbing Milo's whiskers and boxing him with his back legs) were very different. Delilah, our miniature dachshund, and Minnie, our enormous Flemish Giant rabbit, liked to stretch out and groom each other, lolling together like two Odalisques on an old chaise longue.

This win-win relationship extends to interspecies listening too, even, incredibly, by coral larvae and flowers. A 2018 study I can't put down describes how evening primroses responded *within three minutes* to the sound of bees buzzing by producing extra-sweet nectar. Another, equally astounding one describes how coral larvae can be enticed to settle and regenerate degraded reefs by playing them, through hydrophones, the sounds of healthy ones. I didn't realize, I realize, that *coral can hear.*

We would not, of course, know any of this if we weren't listening to their listening.

But why now, when we no longer need horses for battle or dogs for protection or cats for vermin control, should we still try to listen to them? What's *really* in it for us?

Lord Byron touched on one reason in Canto I of his epic *Don Juan*:

> 'Tis sweet to hear the watch-dog's honest bark
> Bay deep-mouth'd welcome as we draw near
>   home;
> 'Tis sweet to know there is an eye will mark
> Our coming, and look brighter when we come.

We listen to our pets because we love them and they make us happy, and ideally, we want them to be happy too. I spend a muddy mid-October indoors at my desk, wading through rafts of scientific studies that show, over and over, that people with pets are both mentally and physically healthier; how animals lower your stress, boost your immune system, and enhance your concentration. Puff wanders in, then Dolly, and settle their chins on my feet. Instantly my ability to tackle this mountain of reading seems to improve.

I read a 2012 study showing that even *remembering* your pet is enough to increase "the number of life goals generated and self-confidence in goal attainment" of human participants. A flush of reports concluding that petting a dog can cause a transient decrease in the petter's heart rate and blood pressure. Another describes how elderly people exposed to fish tanks in Alzheimer's wards display low-ered stress levels. Several show that PTSD-afflicted Army veterans suffer dramatically reduced suicide rates when sharing their lives with a support dog. There's a study that finds pet owners visit the doctor less frequently than non-pet owners, and one to show that time spent with *or even watching videos of* a dog decreases university students' stress levels.

But I find two firm favorites. In second place, a 2001 study of "hypertensive stockbrokers," half of whom were given a pet and half not. Six months later, the paper concludes, the pet-owning stockbro-kers showed lower blood pressure during stressful tasks. I imagine them on the trading floor, yelling, "Sell! Sell!" amid a cacophony of parrots and Irish wolfhounds.

And first prize goes to whoever came up with the idea for the study that proves that among pet-owning couples, both partners will show less reactivity to—and quicker recovery from—a stressful task when in the presence of their pet *than in the presence of each other*.

(There is one 2010 study, though, that describes how twice as many pet owners died within a year of a heart attack as non-pet owners. Perhaps, I consider, it depends how many times their piglets escaped.)

A heady hormonal cocktail is responsible for animals' effect on our well-being: Oxytocin (commonly called the "cuddle hormone"), dopa-mine (prompting feelings of pleasure in everyone from you and me to fish and fruit flies), and mood-boosting serotonin and endorphins all go up, while the stress hormone, cortisol, and respiratory-raising adrenaline go down.

And we're good for them too. "Secure attachment, strong emo-tional bonds and positive interactions between the dog and the owner," finds a 2023 pilot study on PTSD-assistance dogs, "are associated with reduced levels of stress in dogs." Dogs produce oxytocin, say others, when they see a smiling human face and when we look deep into their eyes. And making us happy brings dogs great joy. Simon Gadbois, our daughter Tyger's animal behavior professor at Dalhousie University, told me about Zena, his collie, whom he trained to sniff out rare Eastern ribbon snakes for conservation research. One day, on a field trip out to find them, Simon instead spotted a smooth green snake—*Opheodrys vernalis*, his personal favorite—and picked it up to show his students. Zena watched intently as he waxed lyrical over its red-tipped tongue, its neon-green coloration. *Oh, he likes* that *kind, does he?* she appeared to be thinking. *Well, there's more where that came from.* And for the rest of the day, though never trained to do so, Zena diligently and delightedly sniffed out only those.

Listening to wild animals also works wonders on humans' mental health. A 2022 study found that listening to recorded birdsong reduced healthy participants' anxiety, while hearing traffic noise intensified symptoms in those suffering from depression. The report makes me rise from my researching and go outside to listen to the birds. I plonk myself down on the grass. Instantly, Agnes races over. Her little wet snout presses against my fingers. Her amber eyes gaze deep into mine. *Oxytocin up*, I think. *Cortisol down.* Then she scrambles onto my lap and collapses in a pile of happy wheeps and wheezes.

"Silly sausage," I say. Then, "Sorry."

I feel the most joyous I've felt all month.

I once, while following the daily duties of a veterinarian for a story, had to go to a factory farm. It was a low, airless bunker filled with tens of thousands of identical white chickens. They were crammed into wire-mesh cages, three or four apiece, so tight they could barely turn round, so low they couldn't stand. We walked the aisles, inspecting. On the floor lay delicate, dinosaurish four-toed feet, ripped off from getting stuck in the cages' bottoms. There were bits of comb and beak and wing and eye, where birds had cannibalized one another and themselves. The smell was diabolical. The din, overwhelming. Flies formed a fog so thick I was forced to keep my mouth closed and nostrils pinched shut to avoid accidentally swallowing them.

Our own rescue chickens—Ivy, Sapphire, Marigold, Sparrow, Betty White, Margarita, Tofu, Joy, and Jubilee—were each as distinct and different as our children. Betty White, for instance, was a big, bosomy, buff Orpington, a huge ball of golden feathers with an alert eye that mellowed into an expression of sublime pleasure when she got hold of a blueberry or a grape. Margarita was a pretty Welsummer with lacy brown feathers and almost no comb, who took a pragmatic approach when let out to wander: She was the rear guard, always sending her comrades over the top first.

Still, I could not conceive, under such circumstances, of each of these white chickens as an individual, though I knew theoretically they were. This place was so overwhelming I didn't even feel sad. I felt nothing. I simply wanted, for the sake of my own senses, to escape.

"All's in order," the vet told me, signing the requisite paperwork. This hellscape, it turned out, met all human standards. Evidently nobody had ever asked a chicken.

Factory farms are a modern development. The first ones opened just after World War II. But it's clear that as we moved away from living *with* animals to farming them on this scale, we fast moved away from listening, usually to the detriment of the animals themselves. The day after Agnes looked me deep in the eyes, I read a 2022 report from an EU-sponsored project called SoundWel intended to remedy this, by creating a "tool for farmers to evaluate welfare states of pigs." "Classification of Pig Calls Produced from Birth to Slaughter, According to Their Emotional Valence and Context of Production" runs its snappy title.

Scientists in five research labs across Europe, it explains, recorded seven thousand four hundred vocalizations from 411 pigs. Short grunts, it found, were associated with positive emotions. Screams, squeals, and barks of fear or pain were "of longer duration and more variable in tone." There is also, it says, an expectant sound, like a guinea pig's ascending *Wheee-eeeee*.

Just a month into listening to Constance and Agnes, this is hardly breaking news.

Pigs aren't the only subjects of this sort of research. Tyger calls me from university to tell me a professor there is developing artificial intelligence, or AI, software for listening to chickens in industrial settings. She sends me the press release. "Knowing what these vocalizations mean," it reads, "could deepen our understanding of these animals and improve poultry farming, chicken welfare and quality of life."

Yet as with the piglets, it seems we can do this anyway: A 2024 Queensland University study showed that 71 percent of human participants could correctly distinguish from sound alone when a chicken anticipates a reward. My own experience tallies; over the years, we built up what our children call our chicktionary. We know the squawks of terror when you catch a new hen who has never been handled gently. The purr, quite like Constance's, when she discovers the pleasure of sitting on your lap. Her peaceful crooning when sitting on an egg, albeit unfertilized and never to hatch, and her outrage when evicted from it. The flock's anticipatory clucks in the morning when breakfast and freedom approach; their trills and warbles of contentment as they preen in the morning sun, fluffing their feathers, stretching a wing here, a leg there. The soft, sad song of a chicken as she lies dying, and you stroke her head gently and tell her how much she is loved.

Australian moral philosopher Peter Singer, professor emeritus of bioethics at Princeton, isn't convinced of the benefits, to the farmed animals themselves, of commercial listening paradigms. "Chickens," he says when we talk on Zoom—he in sunny Sydney, me in my rain-soaked anorak—"are the most ill-treated of intensively farmed animals, kept in such great numbers that they don't count as individuals at all. Farmers accept that 5 percent of them will die, despite only rearing them for six to seven weeks after hatching. So once they reduce the mortality rate at a certain stocking density, they'll simply say, 'Let's stock them a little more densely and we'll still get greater profits per enterprise.' That's always the criteria: What's the best profit?"

Peter wrote the classic *The Expanding Circle* in 1981; its tenet, that over the course of human history, people have expanded the circles of beings whose interests they are willing to value similarly to their own. "Listening to animals," he tells me, "is not about profit or yield, but about being the *morally right thing to do*. It's quite parallel to interacting with people from another cultural background. If you asked, 'Why should we listen to the Inuit?' and I answered, 'Well, it would be beneficial to us white North Americans because we'll learn something about ourselves or won't get into unfortunate conflicts that might harm us,' everyone would reject that, because it's obviously the interests of the Inuit themselves that count. It's not a complete parallel, but the point is we shouldn't just be thinking of our own group."

Animals, he says, deserve to have their interests considered, and we can only know what these interests are by listening. This fits in, he explains, with his utilitarianist philosophy.

"I judge actions as right or wrong in accordance with their consequences. A consequentialist ethic says that the right action is the one that will have the best consequences of the options available to the agent. Utilitarianism is a form of consequentialism that says that by 'best consequences,' we mean maximizing the net surplus of happiness over misery."

"Maximizing the net surplus of happiness over misery," I repeat. "I love that. Have you considered," I inquire of him, "owning piglets?"

Marc Bekoff, legendary ethologist, behavioral ecologist, and professor emeritus of evolutionary biology, likes piglets too. He sends me photos of himself with two pigs he knows: Winston and Geraldine. "Redheads rock—truly!" he says when I show him Agnes and Constance. He is not what I imagine scientists to look like; he sports a ponytail and an earring and goes cycling in Lycra, and he's *seventy-eight*. I think of Bernie, bright and breezy at eighty-five. Listening to animals, I conclude, must keep you young.

I first read Marc's classic *The Emotional Lives of Animals* almost twenty years ago. He has recently reissued it and is cowriting a children's book with his friend Jane Goodall, herself a spring chicken at ninety. "On top," he shrugs, self-deprecating, "of my regular ethology type stuff."

Born in Brooklyn, "a kind of concrete jungle," he was, he remembers, always listening to animals. "Talking with them, my folks used to say. I could feel their feelings when I was three, and I still can. I never doubted for a moment that animals could talk and that we must listen. They're constantly telling you stories."

In his early days in a PhD program, Marc named each of the cats his medical teaching laboratory was euthanizing for research and experimentation. "They told me not to, but I did it anyway." Later, he started the first nondissection lab at the University of Colorado, wherein no animals were harmed. Colleagues laughed and said it wouldn't be popular. "But it was the only lab that filled so full they had to add extra sections."

Then for almost a decade, he studied coyotes. "We found there's no such thing as 'the coyote.'" I think instantly of Rachel's robins. "It depends where they live and the food they have and who their coyote friends and families are. A dog isn't a dog isn't a dog. When we were in the field, baby coyotes would come out of the den at about three weeks old. Same hole in the ground, same mommy, same dad. They come out and look around, and there's bold and there's shy and there's some so obnoxious that boy, you wish 'em a good life—but hope you never encounter them as adults. Each one is an individual. And when you focus on the individual first, it's easier to connect to that species."

He discovered the same thing when studying Adélie penguins in Antarctica. "They don't look the same or behave the same. If you spend fourteen hours sitting on your butt watching penguins, you soon get to know their personalities. The fields of cognitive ethology and behavioral ecology really flourished once people started observing the individual."

I think again of those poor, sorry battery chickens. Of course they were individuals. They had just as much personality and consciousness, as many thoughts and feelings and perhaps as much theory of mind as a coyote or an Adélie penguin: We'd simply deprived them of their ability to express it.

"Distancing and separation really open the door to abusing nonhumans. When humans put themselves above and apart from animals, that's a fast track to hell. Listening closely to each individual is the key to preventing it."

The same, Marc says, goes for people. For twenty years, he has taught an animal behavior class to thousands of inmates at the Boulder County Jail.

"Talk about a group that people think of as 'other.' Many folks would feel they have no way of relating to someone who's in a federal prison. But for lots of these guys, as with many of us, it's the animal connection that gets them going. They grew up in such incredibly dysfunctional homes that dogs, cats, monkeys, or snakes were their best friends. Their animals didn't judge them. Their animals trusted them. It was a clean slate between themselves and their pet dog or lizard. They remember that in class, and it helps them open up to compassion and empathy."

Animals listen nonjudgmentally, Marc impresses upon me, in "the now."

"Listening to my incarcerated students that exact same way and teaching them to do the same to both animals *and* humans have helped countless 'othered' individuals I've taught reconnect, heal, and grow."

One Thursday morning at the tail end of October, I wake to find the hummingbirds gone, off on their perilous southern pilgrimage. *Why did they choose last night?* I wonder. Was it the weather? A certain phase of the moon? Magnetite in their miniature hollow bones? Or perhaps a profound, primal urge, like hunger or sleep, impossible to resist?

Will I ever see Mervyn again? I don't know. I unhook the feeder. Scrub it out. Set it down in a cupboard for next year, when I hope they'll come back.

At noon, Tyger calls.

"Mum, our landlord says we can have a cat. Can I take Winnie?"

Though there's been no more bed-wetting, Winnie is not expressing happiness at the farm. She has decided she hates the dogs, whom previously she tolerated. She won't lounge around or play like she used to. She slinks about upstairs, never comes down, then howls at us for company.

"All right," I say. I don't mention the pee.

Next weekend, we drop Winnie off at the apartment Tyger shares with her two university friends. The house smells of perfume and hair products and ramen noodles. There are zero dogs. Immediately, Winnie hops up onto a windowsill to watch city cars go by.

"Hello, Winnie," says Tyger's flatmate, Megan.

In reply, Winnie squeezes her eyes closed, the closest this RBF-afflicted individual ever comes to a smile. Sometimes called a slow blink, the eye squeeze is your cat's way of telling you they feel comfortable in your presence, since closing their eyes makes a feline and even a human immediately vulnerable. "You can," I tell Megan, "eye squeeze back if you like."

"They'll get along purr-fectly," Indy sighs gloomily as we leave the inhabitants of apartment 3B squeezing back and forth at one another across the fairy-lit student living room.

"But it's nice, isn't it," I can't resist responding, "that Winnie's feline fine?"

Back home, I spend another week diving deep into science for more reasons to listen.

I read how interactions between zookeepers and their charges can build bonds that enhance the emotional well-being of zoo-kept primates and carnivores and how just ten minutes of positive keeper-chimpanzee interaction, five times per week, improves affiliative relationships both between chimps and their human visitors and within captive chimp communities themselves. I read a study outlining how canine behavioral euthanasia could be prevented if we better understood a pet dog's undesirable behaviors. How the Save the Whale movement gained momentum thanks to biologist Roger Payne's 1970 *Songs of the Humpback Whale* album, which allowed the turntable-owning public to eavesdrop on hitherto unheard sounds of the deep. The more we listen, the more we know about animals' lives. The more we know, it transpires, the more we tend to care. The more we care, the more we want to listen and know: a sort of beneficial ouroboros.

I chat with the lovely Jonathan Balcombe, ethologist and author of *Pleasurable Kingdom*, which details how animals seek experiences, just as we humans do, simply for feeling good.

"Why listen?" I ask him.

"Because animals aren't just alive. They have lives. They don't just have biology. They have biographies. They have good days and not so good days. Their life matters to them. Listening is an act of empathy. Of caring about them and putting ourselves in their position."

Then I remember my 3:00 a.m. note-taking, *not get stomped ect*, and consider how listening also prevents conflict; how we taught our kids what a bear's body language might mean if we met one out hiking, the difference between bluffs designed to scare you off and genuine threats like huffing, jaw snapping, and paw stomping, and what to do in the unlikely event of an attack. ("If it's brown, lie down. If it's black, fight back," they recited, to which they added gruesomely, "If it's white, say good night.") I recall the mahouts at the Elephant Conservation Center we visited in Laos, who stressed how closely they listen to their twenty-two rescued residents to ensure both human and elephant stay safe. The same goes for horses: Noticing your equine partner's subtle snort or ear flick can be the difference between a relaxing ride and a disastrous bolt from the blue.

I also read how listening to animals in the human quest for knowledge—to understand more, for example, about the workings of birdsong or the evolution of human language—can be ethically suspect. Gloriously, studies have found that mice, both male and female, serenade prospective partners with ultrasonic love songs: tunes far above the range of human hearing that, when slowed down, have the melodic quality of birdsong. But are these songs learned, a group of scientists in 2012 wanted to know, or instinctual? To find out, they surgically deafened mice, then recorded their songs and played them back, slowed down, until they were audible to human ears. Compared to hearing mice, they found, the deaf mice's songs were largely "squawks and screams." Immediately afterward, the report continues, the mice were "sacrificed by decapitation without anesthesia" so their brains could be sliced up and studied, an end to a love story that makes the denouement of *Romeo and Juliet* read like romantic comedy.

Then finally, on Friday afternoon, I encounter one of my favorite reasons for listening of all.

The Swiss Sounding Soil project listens to life subterraneously. Soil, it says, has its own soundtrack: the sound of worms as they muscle through tunnels, of roots as they push down into the earth, the clicks of larvae devouring these root fibers, the thuds and whumps of mole rats' heads and feet against the walls of their tunnels as they communicate with one another. A single cup of dirt, I read, contains many millions of noise-making life-forms, from mites to springtails to centipedes. The project records their sounds with sensors that detect vibrations, which researchers turn into electrical signals and amplify until we can hear them. The end goal: a National Library of Soil Sounds, a sort of underground version of Bernie's biophony.

I watch a video on the Sounding Soil site. One English subtitle captures, perhaps accidentally, its magic: "But the soil is alive," it reads. "Bacteria, algae and fungi cavort in it."

Then I listen to a recording from underneath a cultivated grassland in Mayens de My, Switzerland. It is a file from almost ten years ago, a snapshot of a bit of earth in a five-minute moment in an utterly hidden place. It's enchanting. There are noises like a fax machine, the insistent buzz of a cell-phone message. A creak somewhere between old bones and a distant train. A sort of cooing. A series of scratches. Plenty of bumps, then a trombone slide. Are those distant voices? Is that a cricket? I notice my mind struggling to place the sounds, to conjure pictures to go along. I'm sure that whatever I'm imagining, I am totally wrong.

Listening underground, says Sounding Soil, helps monitor environmental health. It can answer unsolved questions—when, for example, do roots grow most: during the day or at night?—and even work out if the anthrophony is affecting the subterranean world: just as oceanic sound pollution affects marine mammals, noise travels far underground.

But mostly it's just magnificent. All that sound, all that talking, going on everywhere, always. It feels suddenly impossible to ever be lonely—though these sounds are, in a way, the loneliest of all. Listening to that soil talk makes me feel like I did when, with Miss Mander, I watched tadpoles turn into toads. That awe and marvel. The wonder of it all. The heartening camaraderie in how many of us, from plants to pigeons to people, are all listening (even if we don't know what the other is saying). A feeling of connectedness, like Rachel with her robins, in our sometimes isolating world.

The last of the harvest is in. The first frosts have arrived. The summer's cicadas and katydids, whose voices, according to the fossil record, are our planet's oldest, sing their swan song. The piglets cavort in what's left of the pumpkin patch. Ten days before Halloween, Indy, who has been saving up for a Steve Irwin outfit, decides to recycle last year's vampire instead. He uses the cash to buy a motion-detecting trail camera with infrared night vision. "The kind, Mummy," he says, "used by hunters who cheat."

He straps it to a stump, facing a grassy swath between orchard and meadow.

Next morning, we find images of a herd of passing deer. The following morning, a bobcat, with stubby tail and furry ruff, staring straight at the camera. A raccoon mother with her tumbling entourage. A ponderous, clittering porcupine.

Without it, we'd never know they'd been here at all; now we delight in glimpsing, day after day, these night owls' silent forays.

What would happen, though, if one night they all vanished? Would we simply think it weird, or would we take the vanishing seriously? This, I read, is what happened to behavioral ecologist Rachel Grant, who in 2011 was tracking animals high in the Peruvian Andes. For three weeks in August, her trail cam picked up fewer and fewer animals. Then, on August 24, came a major 7.0 magnitude earthquake. This reminds me of stories that surfaced in the wake of the deathly 2004 Indian Ocean tsunami. In Thailand, Sri Lanka, and India, eyewitnesses reported elephants, buffalo, bats, cats, and flamingos fleeing to higher ground, prompted, scientists believe, by vibrational, infrasonic, olfactory, or electromagnetic cues unavailable to our human senses.

So could listening to animals' prescience be to human advantage? I spend Halloween day chasing down the surprisingly few scientific studies to have been published on this subject. I find one French project that sought to track migratory birds (including bar-tailed godwits, bristle-thighed curlews, and a wandering tattler, proving ornithologists possess imaginations second only to namers of English country villages) as potential "early warning systems" for cyclones and typhoons. Another, in earthquake-prone central Italy, that noted "anticipatory patterns" in sheep, dogs, and cows before seismic activity.

But most reports I unearth are either historic or anecdotal or both. A 1908 report by the California State Earthquake Investigation Commission on the devastating San Francisco quake of 1906: "The most common report regarding the behavior of dogs," it reads, "was their howling during the night preceding the earthquake." An account by Aelian, a third-century Roman author, of a tsunami that wiped out the ancient Greek city of Helice: "For five days before Helice disappeared," he wrote, "all the mice and martens and snakes and centipedes and beetles and every other creature of that kind in the town left in a body by the road that leads to Cerynea. And the people of Helice seeing this happen were filled with amazement, but were unable to guess the reason." The night the last animals departed, Aelian relates, the tsunami wiped out Helice and its poor guessers with it.

I'm not sure, though, how much stock I take in Aelian. On the opposite page, he informs us that peacocks carry flax root beneath their wings as protection from the evil eye and that to prevent a frisky mare from going "a-horsing," you must cut her mane to make her feel "downcast at her disgrace."

Perhaps it's because of the wildness of claims like Aelian's that few have been scientifically explored. In a 2005 edition of *National Geographic*, I find an announcement that the World Wildlife Fund intends to study the tsunami stories, but can find no follow-up. The United States Geological Survey—which performed some studies on animal prediction in the '70s, but nothing since—says that though anecdotal evidence abounds, such predictions are fiction: "Changes in animal behavior," it says, "cannot be used to predict earthquakes."

Yet anecdotes aren't worthless. In 2004, with a baby in arms and another on the way, our family lived on the top floor of a city apartment building. One night, Milo began barking on the roof terrace and wouldn't stop. "Shh!" I called. "You'll wake the baby!" But evidently that was what Milo intended to do, dashing into our bedroom, then back out to the terrace to bark some more.

I stumbled out of bed and followed him. "What is it, Milo? Why won't you shut up?"

Behind him, flames were licking the building's sides.

A frantic call to 911. A trip through the smoke-filled lobby, its glass-block windows exploding in the heat. A few minutes more, the fire chief told me, and the bank of propane tanks beneath the building would have ignited, taking all of us with them.

Since then, if the dogs bark at night, I always leap up to investigate. When the horses prick their ears and stare, I stare too. It takes a while before I see the hikers on the old rail trail, or Fräulein and Hildegard emerging from the wood, or a diffident deer passing silently across the pasture. One morning, Dolly sniffs for ages, hackles bristling, around the grass between orchard and meadow. I consult the trail cam. Sure enough, last night, a black bear shambled through.

At 4:00 p.m., Indy appears, face painted, fangs installed, and ready for trick-or-treating.

I glance out the kitchen window. Across the river, the cows have congregated, lying down beneath a tree.

"Take a coat, Nosferatu," I say, "for on top of that cape."

Indy rolls his eyes. Science, he reminds me, says this is an old wives' tale: There's no evidence to support it. He checks the weather app. *Zero chance of precipitation*, it informs us.

I bring our coats anyway.

An hour later, it's pouring.

3

A Round of Pigtionary

How to Listen like a Scientist

> "We have two ears and one mouth, so we should listen more than we say."
>
> —ZENO OF CITIUM, quoted by Diogenes Laërtius in *Lives and Opinions of Eminent Philosophers*

NOVEMBER

Sniff, sniff.

Sniff-sniff-sniff.

I am on my hands and knees crawling around the farm's workshop, sniffing hard, with Puff and Dolly, in all the corners.

Outside, the year has grown grimmer, stormier. It hasn't stopped raining since the cows lay down. Dark mornings and afternoons bookend abridged days. Sven the red squirrel leaps from branch to branch through the disrobing orchard, unwilling to touch the ground. Most snowbirds, on the coattails of the hummers, have flown. On still nights, I hear phalanxes of geese, their honks high and mournful as they follow the river south. But most nights aren't still. The wind's roar has replaced late summer's stridulators: a different cadence depending on whether it slams the house from the west or barrels down the mountain as a furious nor'easter. It's far less appealing now to traipse down to the meadow to listen. When I do, it's pure geophony.

Inside their pigloo, Constance and Agnes snuggle. The horses stand side by side, bums to the wind. Hand-hewn nails the size of asparagus spears start exhuming themselves from the meadow's mud, relics of some long-gone structure.

The night before our sniff-a-thon, I lie down with the dogs beside the kitchen fire and glimpse the house from a dog's-eye view. Down here, I can feel a draft from beneath the kitchen door, see flames dance a shadow cabaret on the shiplap ceiling. I hear the house shift and creak, the scrabble of mice in the hay room beneath us. Puff rolls over and asks for a scratch. I oblige and think of the ingredients required to be a good human listener.

Cicero, 2,500 years ago in his treatise *On Duties*, listed some of them. Take turns, he said. Don't interrupt. Be courteous, and don't talk too much about yourself. "Let there be," he advised, "no exhibition of anger or inordinate desire, of indolence or indifference, or anything of the kind. We must also take the greatest care to show courtesy and consideration toward those with whom we converse." Listen deeply, I add to that checklist. Put yourself in your conversation partner's shoes. Accept that your way of seeing things is not the only one. Try to understand how they think and what they think about. Find out how they feel and why. Pay attention to the details.

I know what I think talking is. I know some of the many reasons to listen. But how, I wonder, do these human how-tos measure up when listening to animals?

I decide to ask three trailblazing scientists whose books I've read and whom I've long admired from afar. Each specializes in listening deeply to one particular species. A familiar canine. A formidable bird. An overlooked rodent. What can I learn from the very different ways these three luminaries approach listening?

Beside the fire, Dolly looks up, suddenly alert. She sniffs for several seconds, then subsides back into sleep.

"Good dog, Dolly."

A slow slap of tail against kitchen floor indicates that despite her slumber, she heard me.

Dogs, science shows, are excellent listeners. They can tell the difference between various human languages and between gibberish and real words. Researchers tested this adorably by reading *The Little Prince* in several languages, both real and scrambled, to eighteen dogs and fascinatingly found that "longer-headed" dogs like Daniff Dolly had "greater auditory sensitivity to speech naturalness" than stubby-faced dogs such as bulldog Puff. Dogs process our tone of voice in one brain region and our words in another and, a 2016 study concluded, show "increased activity in primary reward regions [of the brain] only when both lexical and intonational information were consistent with praise": They know, in other words, that if you coo "Hello, you stupid, smelly pooch," in your cutesiest voice, you're not really being nice.

Dolly and Puff know their own names and each other's. Both know "Sit," "Stay," "Wait," "Fetch," "Lie down," "Head down," "Give me a paw," "The other paw," "Go to your blanket," "Spin," "Touch," "Roll over," "Want a treat?," "Let's go out," "Come back in," "Gently," "Quiet," "Speak" (for Dolly, this command is an imitative "Woo-woo-woo"), "Bedtime," "Walk," and (growled) *Drop that shoe.* If I shout "Bang, bang!" and point a two-finger pistol, both fall down dead like actors in a spaghetti Western. Yesterday, for fun, I taught Puff to "Whisper," at which she wrinkles her bulldog nose and issues her breathiest "uff."

Dogs are also exceptional at reading our body language: They know the difference between us getting up for their leash or to go to the fridge. They can distinguish happy from angry human faces and make more of their own facial expressions, appealing puppy eyes in particular, when they know we're watching.

And studies suggest that dogs' barks originally developed *specifically to talk to us*. Wolves, dogs' direct ancestors, rarely bark after the yips and yelps of cubhood; instead, growls and howls make up the bulk of their oral communication. Scientists think we encouraged dogs to bark during early domestication, since their function was largely to warn us humans about the presence of something nasty or to warn something nasty about the presence of themselves.

Nowadays, though, dogs bark at us for a myriad of reasons: because they want something or don't want something, because they've found something or can't reach something; they bark their chagrin, to their owners' own chagrin, at being left home alone. Science suggests that just as with chickens' anticipatory vocalizations, we humans can automatically understand and correctly categorize aggressive, fearful, or playful barks from unfamiliar dogs.

But what about the subtler differences? Do I study my *Canis familiaris* just as closely as they study me? Have I adjusted *my* means of communication to better hear them? Not really. My dogs know far more Human than I know Dog. Cicero, I think ashamedly, would not approve.

Next morning, to try to remedy this, I talk with the first of my three groundbreaking scientists, Alexandra Horowitz, who for more than twenty years has been listening closely to man's best friend.

She appears on my screen in a plant-filled room with a calico cat behind her and an enormous dog on her lap. This is Quiddity: "a big brindle, possibly part Pyrenees, part Staffordshire, and part who knows."

Quiddity, she says, is slightly needy. "Oh, thank you," she adds as Quiddity trots off and returns with an offering. "You've brought me a jacket."

If you want to listen *to* a dog, Alexandra tells me, you should first understand how to listen *like* one. I consider my listening checklist: *Put yourself in their shoes*, or, in this case, paws.

And dogs, she continues, listen to the world nose-first. "When they're looking at me, they're also *smelling* at me, and that might actually be the first thing they're perceiving." No matter how sharp a dog's eye or ear, she says, it's the nose that is ultimately stronger. "And it's knowledge of the keenness of dogs' olfaction that has to drive my imaginative leap into their umwelt."

"How," I want to know, "should I take that leap?"

"It's a small thing to get down on your hands and knees to the dog's vantage point and see what's visible and audible and smellable down there. But it's another thing to look at their behavior *with* those perceptual organs. For dogs, it's quite explicit. They are sticking their noses *in* things as a way of communicating and receiving communication from others. I need to disassemble my human visual-centered biases and try imagining how the world would be if it were built of olfaction first, not visual stimuli."

I try picturing a landscape made not primarily of sights, but of smells: where a trash can is more meaningful than a sunset, a single orange than a Rembrandt still life.

We humans, Alexandra goes on, could orient successfully around many different environments by smell alone, yet we always default to vision. "We've chosen the big category that we deem most important. But listening to animals is about putting that away."

Human beings, it's true, tend to think of smells as grand, salient things that fit into one of two categories: There are good smells (onions frying, fresh baked bread) and bad smells (dog shit, summer sewers). "But if you put your nose right up close to almost any object like a dog does," says Alexandra, "you'll find it has a 'smell identity'—one that doesn't fit any simple division."

"Isn't that hard, though," I inquire, "when my nose is nothing compared to my dogs'?"

"Don't think of that as an insurmountable barrier," she encourages me. "Think of it as a springboard for scientific discovery. I love the fact that though we're not oriented toward olfaction, we actually have these really *great* noses. Just make some small movements to see how much more you can detect. Your dogs model that for you all the time. Have a go. You might be surprised."

I thank Alexandra, give Quiddity back her human, and invite Puff and Dolly away from the fire to help me try my hand (and knee and nose) at orienting olfactorily.

They burst into the workshop, as always sniffing every surface. The farm's last owner had a Jack Russell terrier. *Can they smell him?* I wonder. *What about further back?* This house was built in 1861: How many of its former inhabitants can they still discern?

I show willing. On my hands and knees, I go where they go. Push my nose, as Alexandra advised, into what they do. They accept me companionably. Wooden crates: sweet, sawdusty. A horsehair-stuffed couch: a mix of grandparents' wardrobe and old, damp dog. Saddles: that comforting smell of worn leather. I sniff a cardboard box; it's had apples in it. A plastic container from the recycling bin. I expect it to smell of nothing, but there's still a savory aroma of Thai takeout, a tang of dishwasher capsule. In one corner, I find an ancient key, teeth all snapped off. It smells blood-rusty. I wonder who used it and where and whether the dogs can tell.

As I crawl about, I remember Howard Lancum's 1954 *Badgers' Year*, which describes this mustelid's astonishing sense of smell. One afternoon, he pressed his hand on the ground outside a badger's sett; when one emerged many hours later, it sniffed, recoiled in horror, and bolted "incontinently" back underground.

But dogs are something else. They can smell whether humans are happy, afraid, sick, or stressed. Simon, Tyger's university professor, is experimentally training a mix of breeds to locate unmarked Acadian graves via a smellable enzyme left by four-hundred-year-old buried human bones. I sniff the metal food bins; I can tell without looking the earthy pig feed from the horses' sultrier beet pulp. I know there's lots I'm missing, but my nose is better than I thought. I imagine the workshop's scents emerging like old nails from mud.

We sniff our way to the workshop's far end, where, behind a little door, there's an old WC with two holes cut companionably in a board above what must have been a night-soil pit. I draw the line at sniffing in here, though Puff and Dolly, I bet, could discern an interesting scatological history.

They seem to enjoy my being with them, shoulder to shoulder. Occasionally, one pauses and gives me an encouraging nudge, as if to say, "*Finally*, you've smelled the light." Sometimes Dolly sniffs at something, looks at me meaningfully, then sniffs again. I take this as an invitation to go in for a sniff myself. Puff's ears, I notice, prick forward when she smells something interesting, but sometimes slide back, flat to her broad skull, as if she's caught a whiff of something strange or unsettling. Her wagging tail stops and droops, her back hunches slightly, then she walks briskly on, as if changing the subject. Meanwhile, I start getting a feel not only for how they might smell the world, but also how they hear it: the thump of each footfall, the squeak of an aged floorboard, the whoosh of the farm's furnace as it flares to life downstairs in the cellar, all far clearer than from a human's perspective up above.

After about half an hour of this, both dogs collapse on the floor. Twenty minutes of sniffing, Alexandra told me, is as tiring for a canine as an hour's physical exercise. No wonder. Because smelling, for them, is time travel: It's like our eyes seeing live-action replays of things that happened months or even centuries ago. *Accept that your way of seeing things is not the only one*: I put a mental checkmark beside another attribute of good listening.

Back at the fire, we form an untidy pile of limbs. Dolly snores up a storm. Puff yelps softly, nose twitching in her sleep. (Can dogs *dream* in smells? Songbirds, I've read, dream in songs, which scientists have recorded via sensors attached harmlessly to finches' syrinxes and played back: Their dreams, it turns out, sound like they're practicing for the next day.) I surf the internet to find out more about our human relationship to smell.

Overall, I read, we consider ourselves desperately un-nasal; most people would relinquish smell over any other sense. But I find an article in *Gastronomica: The Journal for Food Studies* that details how Indigenous and riverine communities in the Peruvian Amazon can smell the shifting scents of river water, thereby sensing environmental change. Water, I read, in its natural state is never the odorless stuff you can buy in a bottle: It always smells of *something*. This triggers a memory. I get up and fetch Edward Abbey's 1968 memoir, *Desert Solitaire: A Season in the Wilderness*. With enough time in the desert, he wrote, a person can learn to recognize the scent of water and the plants associated with it.

Science agrees we smell better than we think we do. Neuroscientist John McGann maintains humans' weak sense of smell is a fallacy created by nineteenth-century anatomist Paul Broca, who noticed our olfactory bulbs are small compared to those of other animals and put this down to evolutionary, "use-it-or-lose-it" selection. In reality, McGann argues, it depends on the topic being sniffed: "Dogs may be better than humans at discriminating the urines on a fire hydrant," he says, "and humans may be better than dogs at discriminating the odors of fine wine. We are," he adds, "about neck-and-neck with canines when it comes to detecting amyl nitrate, the main odorant in banana." Our sense of smell, he concludes, also plays a subconscious role in our encounters with other humans; through it, we can distinguish "kin from non-kin" and discern aggression and anxiety in others. "We even," McGann reports, "appear to unconsciously smell our hands after shaking hands with strangers."

I challenge you ever to shake someone's hand without thinking of that again.

But the best paper I find (though admittedly it evokes a visual, not olfactory, image) is one describing an experiment wherein thirty-two Californian students were blindfolded, earmuffed, and asked to follow, on hands and knees, a chocolate scent trail across a field. Two-thirds succeeded. Four then practiced the task over a two-week period, and all got faster and more accurate.

From this, I glean two things.

First, that there's hope for us scratching the canine umwelt if we just keep on sniffing. I vow to stop and smell more than solely the roses on daily walks, though perhaps not nose to the ground as the dogs do, for fear of raising eyebrows among neighboring farmers.

And second, that Broca's "use-it-or-lose-it" theory applies less to our species and more to the individual.

This takes me back to something Alexandra Horowitz told me earlier. We must always remember, she said, that each dog is an individual and that every individual, within its own species-specific frame of reference, listens and speaks differently. "I put together these large subject groups for an observational study," she said, "but in the end, I almost always want to say, 'But this is just these forty-seven dogs.' And even within them, if we look carefully, we can see so many individual differences that one wonders about the usefulness of even talking about the average of this group of forty-seven as representative of each member."

A cockroach, I think. A coyote. A pigeondividual.

So how does all this help us listen? If we practice listening with our noses, I conclude, besides teasing at the edges of the canine umwelt, we might also pick up on a dog's subtle, individual messages. Because the act of smelling is both an overlooked method of listening more deeply and, of itself, a kind of saying something. "Hey, can you smell that too?" our dog might be saying with their sniffing.

And if it's bananas, our answer is probably yes.

Animal psychologist Irene Pepperberg, my second scientific super-star, knows the value of listening to the individual. But when she walked into a Chicago pet store in 1977 and had an employee choose for her, at random, a year-old African gray parrot she named Alex—short, a little melancholically, for Avian Language Experiment—she hadn't an inkling that listening to him would transform the way the world perceives all birds' ability to communicate with humans and produce the single-most famous avian, barring Big Bird, of the twentieth century.

She's just back from Japan when I catch her on Zoom, having been plenary speaker at the International Bioacoustics Congress. Forty years ago, she tells me, this would never have happened: "All these people standing around, talking about the communication capabilities of a bird with a walnut-sized brain? Unimaginable!"

Funny, glamorous in a ruffled, academic way, and still fascinated by the subject she's been studying for almost fifty years, she is, like most of my favorite scientists, over seventy, but retains the faint Brooklyn timbre of her childhood, when she first got interested in talking birds by way of her pet budgie.

But unlike Alexandra Horowitz, who dives deep into an animal's own umwelt, Irene elected to listen by finding out how much Human one single parrot could understand. In 1977, she came up with a ques-tion: If she taught Alex in a lab setting to speak English, could he learn not just to "parrot," but to respond?

The answer proved a resounding yes.

Alex far surpassed what Irene thought he'd be able to learn. By his death in 2017, he had, incredibly, a vocabulary of over one hundred words, could correctly use numbers up to eight, and had become, Irene relates, the first nonhuman to ask a question in a human language. "What color?" he asked of his own appearance in the mirror, thereby learning the word for, and concept of, *gray*.

He understood other spoken concepts too. Irene could hold up, for example, two keys—one small and metal, the other large and yellow plastic—and ask Alex, "What's different?"

"Color," he could reliably reply.

"What's same?"

"Shape."

"Which bigger?"

"Yellow."

Irene's success in exploring language with Alex came, she tells me, from adapting the learning style of wild parrots to a human-parrot paradigm. "When birds in the wild learn their vocalizations," she tells me, "it's through observational listening. Daddy sings one song, and the guys who are threatening us go away. Mom sings another, then she comes in to feed us. Mom and Dad duet back and forth. And through listening and mimicking and watching our siblings learn too, we babies pick it up."

First, she says, she trained Alex in referential speech: "That is, to identify objects he wanted, then he received that object as his reward. He'd been in a pet store without any toys, and all of a sudden, I was offering him these interesting things: wood to chew, paper to crinkle. So he was very motivated to learn their labels." (To avoid fruitless arguments over linguistics, Irene says she uses a *two-way communication system* instead of *language* and *labels* in place of *words*.)

Each training session saw two humans and Alex alternate through roles of trainer, model, and rival: one asking the questions, the other two vying to get the right answer. "So Alex learned to ask questions and to learn *by* asking questions."

This, I tell Irene, is jaw-dropping.

But to me, Alex's most endearing achievement was what Irene calls "sound play": coming up with his own new words and phrases. After learning the word *talk*, he tried out *chalk*: Researchers showed him actual chalk, and he thereby learned its meaning. Offered a "hatch-day" cake, Alex named it "yummy bread." He christened an apple a "banerry," combining, Irene thinks, the words for *banana* and *cherry*. It reminds me of Indy as a toddler, who, quivering with outrage at some perceived indignity, would launch his topmost insult: "You…you… *meanie bo beanie blah-blah budger*."

Listening to Alex changed humanity's view of what a "birdbrain" is really capable of, and in doing so, Irene checked off another item on my listening list: *Try to understand how your conversation partner thinks and what they think about.*

"The argument," Irene tells me, "was never 'Every gray parrot is going to do this.' The argument was 'A gray parrot *can* do it.'"

Her two current parrots, Athena and Griffin, exemplify the notion that no two individuals learn or *want* to learn the same way. Griffin's vocabulary, she says, is more limited than Alex's, though admittedly he's only had a quarter of the training. Athena, meanwhile, "can talk beautifully when she wants to. But she rarely wants to. In the mornings, when we're getting breakfast, she says an incredible 'green bean,' because she wants her green bean. But then she's like, OK, I've said it; no need to say it again. I think her attitude is 'Life is fine without this human language stuff, so why should I do it?'"

Moreover, says Irene, for his first fifteen years in her lab, Alex lived alone. "So if he wanted to communicate with anyone, it had to be in English. But Griffin and Athena have each other. They can talk in

Parrot. Which is a problem, because why then should they bother with English?"

"And do you understand their parrot-speak?" I ask.

"Well, they go back and forth." Irene shakes her head. "I don't know what they're saying."

This strikes me as rather sad—that we can only understand parrots if they're speaking our language about things we've decided are important to us.

And though Irene's work with Alex remains a spellbinding peek into the capabilities of an avian brain, its focus was far more on what and how much he could think than on what or how he felt. It was indubitably *communication*, but I'm not sure you can call exchanges that largely comprise questions and answers a real *conversation*.

How can we listen to what *feels* important to a parrot, I want to know.

I decide to find someone who can tell me.

Pamela Clark, certified parrot behaviorist, appears on my screen beside Marca, one of her own African grays. It's afternoon for them in Indiana, and she has squeezed me in between back-to-back consultations.

Pam is passionate about parrot welfare.

"I had a husband who was not on board with my profession. He gave me an ultimatum. So I gave him an answer." She flips him, in absentia, the bird. "I moved out with seventeen parrots in a Volvo station wagon."

Pam is who you call if your pet parrot is screeching, biting, self-mutilating, or feather plucking. (Irene, she tells me, consulted her on Alex, who feather-plucked throughout his life.) "These are all common behaviors in captive birds that are basically screams that something's wrong with their environment. That's why, no matter how famous the parrot," she says firmly, "you should listen to them through *their* language, not ours."

Pet parrots, she continues, "*always* tell you how they're feeling about things, and I'm forever saying to people: 'Ask the parrot.' But we tend to assume that verbal language is what matters—that parrots who talk Human are happy. Parrots only talk Human because they're isolated in some way—as a method of making connection."

So really, she tells me, though parrots are considered one of the most vocal of the planet's animals, it's their *body language* that counts.

Save for chickens, I realize, I've never thought much about birds' body language at all. "What kind of body language," I inquire of Pam, "is important to a parrot?"

"For starters, you should *always* look where a parrot is looking to understand where their interest or concern lies. He or she is communicating with you in that moment, but we just aren't listening."

A high priority for parrots, she tells me, is proximity. "These are flock animals. They need to be near others. So when a parrot moves away from something, that's a *huge* communication. But most people acquire animals to meet their own emotional needs, and humans, when they're feeling emotional, want to put their hands on something. Parrots always prefer you don't touch them. But what do we do? We follow them. We pet them. That's how biting problems develop. We keep forcing ourselves upon them, because again, we're just no good at listening."

Grown men, she confides, regularly weep when she instructs them to stop carrying their parrot around on their shoulder.

"Instead, an important form of communication for parrots is parallel activity: In the wild, they bond by eating together, preening together. Create opportunities for parallel activities where you and your parrot are both doing the same thing. Put those mirror neurons to work. Your parrot will thank you—literally."

Parallel activity, like sniffing corners with your dogs, completing a jigsaw puzzle with your family, or playing in a piano duet, is a kind of conversation for people too—that delicious feeling of enjoying a moment with another individual without feeling the need to fill it with active talk.

Not all parrot species' communication, though, says Pam, looks the same. "There are parrots who in the wild live in single-species flocks: African grays, Quakers, cockatoos. Their language is complex and subtle, because they don't have to use broad strokes with each other. But New World parrots tend to live in mixed-species flocks— for example, a family of yellow-collared macaws living with orange-winged Amazons—so their communications are much more overt, because they need to be understood between species."

And how, I ask Pam, does this translate into talking to humans?

If an Amazon is angry with you, she tells me, it's very hard to miss. "That tail is fanned. Those feathers are raised. Those eyes are pinned. Whereas an African gray like Marca is just going to lean slightly forward," she demonstrates, "and subtly raise her shoulder feathers."

Amazons, she continues, are a hoot. "You're going to put on music and dance and scream and do a lot of training, and you'll always know where you stand. But African grays are chess players and negotiators—I know this sounds anthropomorphic, but your relationship with a gray is going to be much more intimate."

On top of that, she says, you layer a parrot's individual personality, demeanor, and preferences.

"Each parrot will have these special little things they do to interact with you. I might walk by, and Marca will make a funny sound. I'll turn and laugh, of course, and she understands that, and it builds rapport. It's all about understanding their priorities, then watching them closely, first as a species, then as an individual. I think *you* should have a parrot," Pam adds as she gears up for her next consultation. "You would love an African gray."

I hesitate. "A dog is for life," the saying goes, "not just for Christmas," but parrots take this to another level. I think of the great-aunts in my family who inherited a saucy old bird from their own great-aunt. Cookie, I've read, the oldest Guinness-World-Record-certified parrot, lived to "at least eighty-two." Lorenz's cockatoo in *King Solomon's Ring* once bit off all his father's suit buttons, trousers included, while he napped in the midday sun. I don't generally have a fear of commitment. Horses can live over thirty years, pigs up to twenty. Husbands, if well tended, might make it to one hundred, and children, we hope, we'll have for life. But I already spend a great deal of time listening to the feelings of my nearest and dearest. I'm not sure I've room in my own birdbrain for a parrot's feelings too.

If I bring one home, I tell Pam, it might well be my family flipping me the bird.

It's November 12, and behind the fence, Dolores the groundhog has gone into hibernation, retreating deep into her burrow, where her heartbeats will drop to just five per minute, her body temperature from 98°F to 38°F. We won't glimpse her again until February at least.

"I hope she won't be lonely," worries Indy.

She won't, I assure him. Groundhogs are the introverts of the rodent world, the very opposite of their community-minded cousins, whom we once encountered on a road trip in the wilds of South Dakota.

It was the sort of middle-of-nowhere rest stop that encapsulates everything cinematic about the American Midwest. The sky was a cobalt dome, the highway a black ribbon laced into the far yellow distance. An old lady named Pearl, in horn-rimmed glasses, was serving free coffee.

We piled out of the car. We stretched and blinked. Then a movement in the scrub nearby caught my eye.

"What was that?"

"It's a prairie dog," gasped Cassidy. "No—wait. There's two of them. No, four. No, look! They're *everywhere*!"

Sure enough, behind the rest stop lay a prairie-dog town, with holes and hummocks and mothers and babies and prairie-dog sentinels. We stayed for almost an hour, watching a world every bit as social and industrious as the Richard Scarry Busytown books the kids pored over in the car.

But prairie dogs suffer from an image problem. To the untrained eye, including mine, they all look the same: small, brown, fat, fuzzy. And they've historically fallen prey to earwig-level bad press. Ranchers say cattle break their legs in their holes. They are shot, trapped, bombed, poisoned, electrocuted, buried alive, and run over. Animal behaviorist Con Slobodchikoff, my third stellar scientist, has spent his career striving to change this, by constructing his "prairie-dog lexicon."

I connect with Con, another bright, brisk septuagenarian, on a particularly dark and gloomy Monday morning.

"When talking about my life's work," he begins, "if I start off with a conservation message, saying prairie dogs are declining, that we now have only 1 to 2 percent of the number we had one hundred years ago, people's eyes glaze over because they've heard this so many times about so many different animals. But when I say prairie dogs *talk*—that they convey meaningful information to one another about significant, specific things—'Oh,' they say, 'they're kind of like us. Maybe we shouldn't be killing them.'"

Con approached thirty years of field-research listening using another of my "good listener" criteria: by *paying attention to the details*. Though to the uninitiated, prairie dogs' alarm calls, for instance, might all sound like the same cute, urgent little *Eep! Eep! Eep!*, he found they're in fact complex, grammatical, and brimful of specific information. They have separate calls for an approaching coyote, domestic dog, hawk, and human, which elicit different, predator-appropriate responses from the prairie-dog community. But that's not all, he says; their calls contain descriptors: specific adjectival and adverbial "words" for a human wearing a blue T-shirt, for example, versus one wearing yellow. They can describe, Con discovered, sizes, shapes, the speed at which a predator is approaching, and the danger level posed by that predator: he found, he tells me, that the prairie dogs' call for his blue-shirted research assistant changed after that assistant fired a gun into the air.

So subtle and elaborate were Con's findings that some scientists (the mentophobes and anthropodenialists, I think but don't say) refused to believe they were real. Why, they queried, would a prairie dog communicate such specific information—say, the person in the yellow shirt versus the one in blue—when the recipient of that warning flees regardless?

"I say they put that in there because it's relevant to *them*: Predators have different coat colors. There are no blue coyotes out there, but there are yellower coyotes and browner coyotes, and it's relevant to know who the individual is based on that coat color. Animals have language systems designed to fit their needs, just as we do. That information matters to *them*."

That "absence of evidence" thing again. Just because we don't know *why* they do it doesn't mean they *don't*.

But Con's research, though captivating, is based on the close study of a narrow slice of prairie-dog talk.

"I picked alarm calls," he tells me, "because they elicit a complete context-response sequence. But their communication is much deeper than we've been able to scratch. For instance, there are many visual signals, plus their social chatters, which we haven't so far been able to decipher. So we don't yet know whether a call like 'Chitter chatter chitter' means 'Do you know where Joe was last night?' or just 'Chitter chatter chitter.'"

And where Irene Pepperberg was interested in what parrots say in Human to humans, Con focused solely on what prairie dogs say in Prairie Dog to others prairie dogs. But what's said, I want to know, when prairie dogs and humans share space?

I track down one last, lesser-known expert, Dr. Gena Seaberg, who can tell me.

Gena has worked with well over one hundred thousand prairie dogs, many of them pets in captive settings, and believes, like Pam of parrots, there's as much to be learned from their body language as from their vocalizations.

"Oh, they'll tell you loud and clear what's going on," she affirms when I call her after lunch. "But you have to learn to read it."

To have a meaningful human-prairie dog conversation, she says, "you need to step out of being that human. You have to get past your preconceived ideas of what the world means and how it looks."

The first thing to understand is that pet prairie dogs "truly see every household member as a prairie dog too, with jobs to do to survive the day. I know some who'll chase a bull mastiff around the house with no trepidation at all if they think he's not pulling his weight. They really don't see them as different from themselves. So they'll talk to you like one of their own."

Some of the gestures important to this egalitarian little animal, she tells me, are ones we humans already understand well.

"They have a 'shunning' behavior which I compare to being in middle school, when your friend is mad at you and acts like you're not there. Prairie dogs will do that to their owners, and it's a powerful thing to do back to them when they're acting poorly."

But other, more subtle signals, she says, are easy to misinterpret.

"What you might perceive as a perky or eager look can actually be they're on high alert for things that might harm them. And when they open their mouths wide—*Eeeee!*—in a toothy smile, it's more than a greeting; it's an *assessment*. They're smelling you for adrenaline, and based on that, they'll decide how they're going to interact. So if you're anxious, hurried, worried, won the lottery, they don't know the difference. They just smell in you chemically that something's off: 'I see you're excited,' they'll tell you, 'I don't know why, and I don't want to touch you or you to touch me.'"

Back in the verbal realm, she tells me, prairie dogs have dialects: an important consideration when relocating wild populations from areas of human development. "We're careful mixing rural with urban prairie dogs, because they have different 'words' for things. They'll look at one another, like '*What'd* you say?' or 'I'm telling you to run for cover, but you don't understand.'"

In a human home environment, they display dozens of other distinct communications.

"If they don't want you to handle them, they'll puff up and rattle like rattlesnakes. Other times, they'll tell you off by chattering at you or issuing low, guttural growls. If they're upset with you, they won't make eye contact. And they'll vocalize with a 'yahoo' or 'jump-yip' to show they're excited: Owners come home and think their prairie dog is happy to see them, but it's more a 'You disappeared from the colony all day, and I'm relieved you're still alive.'"

All this, she concludes, makes for a pet who, though hard work, is nonpareil. "Imagine everything you want out of your relationship with your closest family member, then amplify it: That's what these guys bring. The bond is kind of indescribable. If you find the right prairie dog, it wants to be with you so much that if you could be sewn together, that would be just fine. They'll sleep in bed with you every night. I was bedridden during my first pregnancy, and my first, special prairie dog wouldn't leave me. That," she admits, "was how it all started."

Given all this incredible communication, it seems tragic, I say, that prairie dogs go so unlauded.

"Some ranchers," Gena replies, "have realized they're good for fire suppression, for nitrogen-rich soil, for better grasslands. But is there the type of explosive education where we find the prairie dog's the hero? No. We still see them as pests, even though their numbers are very low, even though they're a keystone species, contributing so much to so many."

And even though, I add, they've got so much to say.

"But maybe it'll change?" I suggest. "With your work and with Con's?"

"Maybe." She shrugs and sighs, her message crystal clear. "Maybe."

Dogs. Parrots. Prairie dogs.

The species. The individual. The umwelt.

These things are on my mind next morning when, setting out our recycling curbside, I find myself staring at the asphalt. Those dotted yellow lines, I reflect, those symbols of safety I see with my trichromatic vision, *are not real.* Or at least they're just my one sub-jective experience of reality. Fifty yards away, our neighbor's tabby cat crosses the road. He doesn't discern these yellow lines' meaning, or even see that yellow with his dichromatic vision. Rather he sees them, like himself, in shades of gray. And that starling up in the oak, a tetrachromat who sees polarized light we can't: For him, those lines and the road and the orchard, all of which he perceives with his 340-degree vision, are disco shades of ultraviolet. And the coyotes, when they trot this way from wood to wood tonight, won't bother with the lines at all, but will see in scents, back through history, the starling and the cat and me, standing here, with my bags of recycling and horse shit on my shoe.

I once read that English speakers picture the future before them and the past behind or as a timeline drawn from left to right. But Mandarin speakers conceptualize time headed downward into the future, more like a family tree. Do nose-first canines, I wonder, experience time as an olfactory ocean in which everything floats, some moments near, others farther away?

"Thus we ultimately reach the conclusion," wrote Uexküll, whose 1934 essay, "A Stroll Through the Worlds of Animals and Men," scrambled my mind this morning as I drank my first coffee, "that each subject lives in a world composed of subjective realities alone, and that even the Umwelten themselves represent only subjective realities."

I could get lost in this all day: in the notion that we're all so different that there is no common ground for communication at all.

Instead I drop the recycling, remember Alexandra Horowitz's springboard, and have an idea.

Con has his prairie dog lexicon. I've browsed an Elephant Ethogram, which comprises hundreds of searchable pachyderm communications, labeled lovely '80s-band-named things like "Alarmed Trumpet" and "Let's-Go-Rumble." I've already got an ad hoc chicktionary. Despite all Uexküll's unweltian mind bending, I also have some listening tools to work with.

I will, I resolve, construct a pigtionary.

Like dogs, Constance and Agnes are listening closely to our human world. At just four months old, they know that kitchen lights at 7:00 a.m. mean breakfast is imminent. That the crunch of tires on our gravel driveway means somebody's home and is heading their way. That dinner held in bowls above their heads means "Stop jumping, you beasts, and I'll feed you."

Now, as the weather worsens, I devote a shivering week to listening back. I suit up and camp out in their enclosure to combine what I've learned from science: I think, like Irene Pepperberg, about what they're thinking about. I consider, like Pam Clark and Gena Seaberg, how they're feeling. I watch for nuances like Con Slobodchikoff. Sometimes I observe. Sometimes I interact. They talk Pig to each other, and they talk Pig to me. Occasionally, I experiment, unsuccessfully, with making pig sounds of my own. Like Alexandra Horowitz, I try to put myself in their trotters: their eyesight (poor), their hearing (good), their sense of smell (impeccable). I show an interest in the things they're interested in. Agnes likes to housekeep; when I give them fresh hay, she plumps and puffs it like a stylist with scatter cushions. Constance enjoys sleeping on her side once the housework's done. I join in with both: parallel activities, as Pam prescribes her weepy parrot men. The piglets seem to like that too.

I also catalog their sounds. No SoundWel squeals of pain or terror here; these girls haven't known a moment of misery in their little ginger lives. But excitement, anticipation, pleasure, frustration, irritation, desire, and curiosity are all expressed to me and each other through two dozen subtle varieties of squeal, grunt, oink, bark, pant, and wheeze. I run in for hot tea or to change my soggy clothes. As soon as they hear me coming back, they start to *wheeeeep*. As I get closer, these become staccato grunts. *Hunk, hunk, hunk-hunk, hunk, hunk...* They are saying, I'm quite certain, they are happy I've returned.

I record them and play them back slower. Do they make a different sound when I appear than when Indy does? Are there descriptors, or words, in all that grunting? I'm not good enough at this to tell. I do discern, though, two very different types of *wheeee*: the excited "Are you coming? Faster! Faster!" and a more curious "Is there somebody there?" when I approach in the dark or from the garage, an unusual direction.

I also watch their body language closely. Sidling up with one shoulder means "Pet me, please." Agnes still wags her tail when she's delighted, whereas Constance raises her snout from side to side. I read a 2018 study confirming that "high-duration tail movement is an indicator of positive emotions," which also says that flattened ears, as with dogs, horses, and apparently sheep, connote negative ones. I see no ear flattening, though, not even when they squabble over food.

There's no such thing as "the piglet," I conclude. This is all just a snapshot: of these two Kunekunes on these seven days, in this sort of weather, at this particular age, here with me in the November sodding cold.

Then on the seventh day, my birthday, there's respite. A thin sun comes out, and I let them out to graze the last fall grass. I listen to their rhythmic, contented oinks. Constance heads off around the corner in search of greener pasture. Agnes grazes close to me, snout snuffling among the fallen leaves. After a minute or two, Constance's oinks rise in pitch, volume, urgency; she has strayed too far and realized she's all alone. "Help! Where are you?" she is calling. She comes galloping back, tiny legs flying, a little winged Pigasus.

Agnes barely looks up as her sister crashes to a halt.

Then Constance makes another noise, a slightly higher, sighing sort of oink: relief, I think. She nudges my hand to ask to be petted and plops her bottom down while I stroke behind her ears, along her back, down her glowing amber sides. Slowly, she capsizes. She starts to purr.

This little sequence reminds me of a time in Manhattan when the children were small, and Zeyah, aged about four, left the group to sally farther into the depths of Central Park's massive Heckscher Playground. There he played, contented, for a while before discovering he was all alone. A moment of panic. A dash back to his siblings, then to me for reassurance.

Constance, I realize, was just a worried four-year-old in New York City.

Most of what I hear and see over these seven soaking days I need no scientific study to verify. I just sort of *know* it, the way the chicken guessers and dog listeners in the experiments just sort of *knew* what the calls signified. I wonder if this has to do with something called the motivational structure hypothesis, which I read about in an almost-fifty-year-old research paper—that there is indeed a universal language out there, based not on finicky words or syntax or grammar, but on *tones*.

The idea, said its author Eugene Morton, stems from Darwin's principle of antithesis, the theory that it's evolutionarily advantageous for an animal to express opposite emotions in correspondingly antithetical ways. In other words, deep, harsh sounds across many species will always signify hostility, whereas higher sounds will designate friendliness, appeasement, or fear. This most often applies, Morton found, when animals are physically close together. Add distance, and because of the way sound carries, the rules may change. This hypothesis isn't a catchall: There has been plenty of research since into cases that don't fit the rule. But still, there might be some truth to it. The ascendant, anticipatory sounds of a pig, chicken, and guinea pig definitely have some commonality. Lots of species, our own included, squeal when they're hurt or frightened and growl when they are threatening someone. And maybe that's what we humans are sensitive to: the cases that fit Morton's motivational structure, a sort of animal Esperanto.

As the November sun sets red and weary, I go out to feed the piglets dinner. All was fine when I put them away this afternoon. But now Agnes's little orange bottom is a mess of slimy green poop, her corkscrew tail weighed down straight. Poop drips from her like an unclosed tap. There are trails of it across the pigpen.

"Oh, Agnes!" I cry. "What happened? What's wrong?"

But even Alex the parrot could not answer questions like these.

Instead, head down, she offers a small, quiet, high-pitched grunt, turns her body away in the familiar manner of a downcast dog, and drips some more. She is telling me, quite clearly, that she doesn't feel tip-top. I watch closely. Her tail isn't wagging. Her ears, for the first time, lie flat against her umber skull. Yet these signals aren't enough to tell me what caused the problem or what I should do now.

She gives me another sad little oink and goes off to lie down alone in a corner.

In the scheme of things, this is no great shakes. An inconceivable four million pigs, each an individual like Constance or Agnes, are slaughtered worldwide per day, and most of our girls' littermates are being fattened for next year's bacon. But I love these little piglets. As Jonathan Balcombe said, they aren't just alive, they *have* lives. And it's precisely now I remember how stressful it is to have a sick baby. Human, kitten, canine, or, it turns out, porcine, we fret and we fuss and, most unhelpful of all, we google.

Never google "piglet with diarrhea." According to the internet, there is no outcome other than certain death.

"Isn't there a Pig MD?" asks Indy.

"She probably just ate too much grass," says Gal.

"Oh, but what if she found a poisonous shrub?" I wail. "What if she gets dehydrated? What if I've accidentally *killed* her on my *birthday?*"

Think, I tell myself. *Think, think, think.*

I text her breeder.

u cld try ACV, comes the reply, n ive heard canned pumpkins gd 4 the runnies but its nvr happpened 2 me so not sure gd luck. She concludes with a smiley face and a shamrock.

All right. Apple cider vinegar. Canned pumpkin. And perhaps the addition of a birthday-cake wish.

I prepare Agnes's dinner.

"Would you like some of this?" I offer.

I hold my breath and try to listen really, *really* carefully.

I watch her sniff the food, then tentatively try a bite. She looks up and gives me the tiniest tail wag.

"Yes please," I think she's saying.

She eats. She drips. She goes off, forlorn, to her pigloo.

I get no sleep on my birthday night.

Next morning, she seems a little better, though her grunts remain soft and sad. I try to imagine how she's feeling, what she might like, and bring her fresh hay, which, half-heartedly, she rearranges. Her diarrhea has turned black. I stick to the diet, and to Constance's disgust, she does too.

Day three sees an improvement. No diarrhea and her sounds are brighter, her body language perkier. I can sleep the night again, except when, predawn, I hear the coyotes chatting nearby and am compelled to check they're not planning a bacon breakfast.

By day six, Agnes is back to her usual piglet self, voicing once more her joy, her excitement, her frustrations, annoyances, and unequivocal demands.

I'm glad I have been listening closely enough to know.

Then on November 29, a storm hits us hard. The house creaks. Its casement windows rattle. I remind myself it has stood here since Victoria was on the throne.

The piglets haven't, though, and at 2:00 a.m., I can stand it no more.

Pajamaed, I slide my bare feet into boots I accidentally left outside, which now contain an inch of cold water. The wind slams me side-on. I wince at the sting of each raindrop as I struggle to the pigloo and shine a light inside.

"Girls?" I call. "Girls?"

Nothing. Then two snouts appear from the hay. Grunt, grunt. Grunt, grunt: soft, inquisitive.

"Who goes there?" they are saying. "Is everything all right?"

In this moment, our collective senses largely useless, our umwelts seem closer. We're just reaching out to each other through the darkness into the howling void.

"Sorry, sorry!" I yell. "Just checking. Go back to sleep!"

The snouts withdraw.

I have been using my human communication, I reflect as I am rained on and buffeted back to bed, and they their porcine one. Like a scientist with a groundhog, I know a bit of theirs, and like dogs and parrots, they know a bit of mine. We're not elegant, like Irene's language-learning experiments. Not precise, like Con's deep analysis of alarm calls. But there's a pinch of body language. A dash of fifty-year-old motivational structure. A determined effort, like Alexandra's, to dismantle that species divide and a mutual willingness to meet in the middle.

Not bad, I think, for a forty-six-year-old novice pig owner and two four-month-old novice pigs.

4
Still Waters

How to Hear the Sounds of Silence

"To me, nothing is voiceless."

—SIMONA KOSSAK

"Silence is of different kinds, and breathes different meanings."

—CHARLOTTE BRONTË, *Villette*

DECEMBER

Winter arrives. The valley turns sepia: endless, dreary shades of yellow, beige, and gray. Frost piques the ground each morning. A foot of mud is now topped with a momentary, deceitful crust. Days are so short that you're barely easing into one before it's dark all over again.

The horses have grown fluffy winter coats, the piglets too: Thermoregulatory, these long hairs trap a layer of air, providing insulation. The world, meanwhile, is quieter. Starlings sing a contracted dawn chorus. Sven the rattling squirrel dozes his days away in a liminal world between waking and sleep. At night, I hear scuttles in the walls, but never a squeak. One mouse makes it as far as the kitchen, though, and I find him next morning gazing up at me, a chestnut roly-poly, having gorged himself silly inside an open sack of dog food.

Outside, only the occasional crow passes over, its wings sawing a dry, squeaky path, and Siegfried's cries pierce the December fogs like knives through paper.

The silence makes me wonder. What about those creatures who are *always* quiet, who don't chitter like chipmunks or chatter like parrots? Those whose voices are outside our human auditory range or too unlike our own to understand, those with whom we have literally no common ground? What about those we've always thought of as silent and solitary, too removed from humans to have any interest in talking to them? How can you listen to *them*?

To try to find out, I travel four hundred miles south of home to Boston, Massachusetts, where skies are blue. It's Monday morning, almost nine, and near the Greenway Carousel, upon which children ride lobsters, crickets, and a soaring peregrine falcon, the New England Aquarium is sloshing to life. Harbor seals peep out at a fizzing stream of schoolchildren. A quartet of young military men in cocked hats and shiny shoes gaze in at a tankful of sand-colored skate.

But upstairs, behind the scenes of the Northern Waters gallery, the activity is of a more urgent nature. Barbara the Atlantic cod, a glistening three-foot female, is being readied for a spay operation, having had trouble releasing her eggs.

Currently she's lurking at the far end of her Sandy Bottom habitat.

"She definitely knows something's afoot," Nicole, an aquarium volunteer, tells me. "They're way smarter than you might think."

It takes a while for Nicole and aquarist Jordan Baker to get Barbara netted and into a trolley marked "FISHES" for the journey upstairs to the aquarium's medical center.

"Good girl," coos Nicole as Barbara thrashes her indignation. "That wasn't so bad now, was it?"

"What's in there?" a maintenance man in the service elevator wants to know.

"She's a cod."

"I love cod," he grins.

Nicole gives him the fisheye. "It's all right, Barbara," she soothes. "You're doing great."

Once Barbara is safely deposited (10:00 a.m. COD SURGERY reads a note on the veterinary whiteboard), we head back downstairs to where a thrumming maze of complex life-support systems regulate, salinate, and filtrate the waters of the aquarium's thousands of silent residents for whom aquarists are busy preparing food and enrichment activities.

Five American lobsters, in litmus hues from hot pink to peacock blue, repose thoughtfully in tanks.

"You might not think so, but they each have personalities and emotions," says Megan, an intern who is busy siphoning one tank clean. "There are the sweethearts and the grumpier ones. This one"—she points—"is very pinchy with the siphon, especially if I'm cleaning around her flowerpot. She'll grab hold of it and won't let go. I just have to let her have it and come back later."

The same, she says, goes for the half dozen redfish next door. "I know how each one likes to eat. Some are picky. Some tuck in with gusto. That one spits out her vitamins like a naughty child."

This reminds me, I tell her, of a study I read recently on archerfish, those beguiling black-and-white tropical fish who spit jets of water to catch their prey and who can be trained to recognize human faces.

"Oh, definitely. There's much more to fish than we see on the surface."

We draw up at Sedna, the grande dame I have come to meet, who is resting in a corner of her tank.

Sedna is a giant Pacific octopus, ten feet long and forty pounds in weight, named after an Inuit ocean goddess and, at around four years old, not the frisky young thing she once was. She is slowing down, showing signs of end-of-life senescence; incredibly, even the largest octopus species don't live much longer than five years. She spends more time asleep, arms coiled in her tank's many crevices, her breathing, at eight times per minute, slow and regular. She is, says Jordan, a reserved, gentle soul. "We've had others," she tells me, "who've been really large and in charge."

Sedna's interests include eating live crabs and mussels and playing with puzzles and toys. "There's a particular red ball she'll hold on to all day—sometimes for ten hours. That's her favorite."

This morning, to my delight, Jordan is letting me spend time with Sedna in the hopes of learning how an octopus, a wholly silent creature with no vocal signals, familiar mammalian traits, or affiliative bonds with its own or other species, might communicate with a human being. As instructed, I plunge my hand into her chilly waters and splash about: a sign Jordan trains every octopus to understand means "I've got something that you'll like. Come say hello. Come connect. Come see."

"The octopus," wrote Aristotle, "is a stupid creature, for it will approach a man's hand if it be lowered in the water."

Sedna awakes, eyes my hand, and doesn't stir.

I splash again.

Nothing.

"Am I doing it wrong?"

Jordan grins. "Sedna's very particular. She often stiffs me too. But Nicole has no problem. She must be speaking her language in a different way. Here. Try this." She skewers a piece of thawed frozen shrimp and hands me the stick.

I hold it out to Sedna, wave it about a bit. She watches, then, with seeming reluctance, accepts my offering. She also keeps hold of the stick. I pull. She pulls back: surprisingly stolid, muscular, insistent.

From beneath us, delighted squeals erupt, and I remember there are visitors down there at aquarium level, watching her every move.

So here we are then: a tenuous moment of connection.

Gingerly, I stretch my other hand out toward her. Mirroring me, she curls another arm out too. It unfurls tranquilly, closer and closer, thinner and thinner, stretching, stretching…until the topmost tips of her suckers are less than a half inch from my fingertips. I reach out, I can't help it, to touch. Too quick. She snatches it back, quite literally recoils.

Her message is clear. *Get lost*, she's saying. *You've overstepped my mark.*

I think ashamedly of Pam's weeping-men's parrots and of Andvari back home, who when he first arrived took months to warm to the notion of my petting him. Touch, I know full well, is earned. How would I feel if some stranger made a grab for *my* suckers? Why should Sedna care if that stranger was only clumsily seeking connection?

Instantaneously, Sedna unskewers her lunch, releases the stick, and flushes from Gothic purple to violent crimson at the taste of shrimp. She pinwheels her arms gorgeously, fluidly, one over the other, a living netsuke, shedding the tips of her suckers, the chitinous cuticles, into the water. They spin off like miniature moon jellies to settle on the floor of her tank.

What is Sedna thinking as each sucker surfaces, then dives back in? Does it feel good, this early-morning stretch of her limbs? Is this the end of our conversation or just a change of subject? With six hundred million evolutionary years since we all inhabited the same flatworm body, I can't even begin to guess. Nevertheless I sense…*something*. An ocean of feeling beneath that soft, silent, color-shifting surface.

"Jordan? Can you come help me?"

Jordan hastens off to an intern's aid.

For the next hour, we stay together, Sedna and I, amid the cool, dark hum of the gallery. She moves closer. I know she knows I'm here. I see her seeing me. She sees me seeing her seeing me. Our eyes meet. I don't try to touch; I just sit beside her, exactly as I did in the early weeks with poor, fearful Andvari, and just *be*. It's a feeling I recognize—apart, but together. Separate, but connected. Oxytocin and serotonin up, cortisol down. Of parallel activity and proximity, in this shared moment, being enough.*

You don't have to fill a gap, I have discovered, with talk or touch, those very human impulses, to be in conversation with an octopus. This reminds me of Thoreau. "What sort of space is that which separates a man from his fellows and makes him solitary?" he asks in *Walden*. "I have found no exertion of the legs can bring two minds much nearer to one another."

Back home, I arrange a call with Dr. Ahmad Abdella, Instagram's warm, charismatic "octopus whisperer," to find out if this moment of togetherness was just wild wishful thinking.

Not at all, Ahmad tells me. "If you're standing eye to eye with an octopus, they really are speaking to you through their gaze. Every time we've locked eyes, I've felt a deep communication and connection."

So powerful is this feeling, he says, that it has inspired him to rescue octopuses from ill-equipped pet stores and seafood markets, where he finds this naturally solitary creature packed in with dozens of others.

How much, I ask him, can an octopus communicate to or with a human?

In answer, he tells me about Byrdie.

Byrdie was a rescued bimac, or California two-spot octopus. Bimacs, unlike Sedna, grow to just seven scant inches and in the wild live about a year and a half.

---

* In 2018, researchers at Johns Hopkins University dosed a male bimac octopus with MDMA, otherwise known as Ecstasy, which prompts feelings of connectedness and elation in humans by causing the user's body to release large amounts of serotonin. To scientists' surprise, the MDMA had a similar behavioral result in the octopus as in a human Ecstasy taker, enhancing "acute pro-social behaviors in Octopus bimaculoides." The result, in other words, was an exceptionally cuddly individual.

"When I saved her, she was missing four arms and had a big gash on her mantle. Nobody thought she was going to make it. She looked half-dead. She was very scared and very small. It took weeks of me sitting in front of her tank, drinking my coffee, eating my sandwich, day in, day out, until finally she started to come out and eat and trust me."

Eventually, though, she sought out Ahmad's company. "I would put my hand in the tank, and she'd come sit right on it." She grew bolder, more curious. She started playing with toys. "At first, she didn't know how to unscrew a jar lid. But after a couple of months, she was opening the safety caps on pill bottles, and that's a pretty hard cap to do."

She also became more opinionated. "It got to the point where she would throw foods back at me that she didn't like. If anybody else came up to the tank, she'd squirt water at them. She would wave people away. But not me. We were very close."

Byrdie lived with Ahmad for almost three years—a lifespan, he says, unheard of for her species. During this time, she charmed not only Ahmad and his family, but also tens of thousands of social media followers.

"So many people stopped eating octopuses because of Byrdie. And others told me, 'First I started feeling bad about eating octopus. Now I feel the same about chicken nuggets or hamburgers.' They realize there's an individual in there. It's not an *it*, it's a *who*."

So much *who* that her death, though expected, was devastating.

"She started slowing down, sleeping more. In my head, I was totally prepared. But when it happened, it was like a bag of bricks to the face." Ahmad sighs. "I miss her."

Byrdie's legacy, he says, is the small sanctuary he established to rescue other octopuses in dire straits, each one its own person.

"We can't know what it's like to be an octopus. To taste through our suckers. To see color in light waves. But there's nothing wrong with relating some of their behaviors to our own understanding of being human. 'Do octopuses like to eat snails?' someone will ask me. And I say: *It depends*. Each octopus has its own preferences, its own characteristics. Some, for instance, always use one particular arm to eat with: like being left- or right-handed. Each one's an individual who it takes time to get to know. Just like us."

Just like Barbara, the cod. I drop Jordan a line to ask how she's doing. Her operation was a success, Jordan tells me. She is back, no longer lurking, in her Sandy Bottom habitat.

I'm still thinking of Sedna a week later when, in weather that bounces between rainstorm, windstorm, hailstorm, and back again, I walk the dogs, cuddle the pigs, hang around, fingers frozen, with the horses. These are all social animals. They thrive in company; like humans, their evolutionary survival has depended on it. But what intrigues me is that even old-lady octopuses seem willing—interested, even—to interact with humans when in the wild, these silent, solitary creatures barely connect with one another. Is it another latent capacity, I wonder, like prairie dogs possessing adjectives?

I pose this conundrum to Australian cephalopod scientist Alex Schnell. It's nighttime, and just in from the evening feed, I am bundled up in several sweaters; it's morning in Australia, and Alex is wearing running clothes: two humans communicating information about themselves, though nowhere near as spectacularly as the creatures she studies. (A male cuttlefish, she tells me, can flash courtship colors to the ladies on one side of his body while simultaneously displaying female colors to fool potential rivals on the other.)

Alex doesn't know the answer. But she has been intrigued by this question since she met her first octopus at age five in a rock pool at Clovelly, New South Wales.

"I saw two little eyes looking out at me, and when I reached my hand toward them, a small arm reached back and wrapped around my finger. At first, I was taken aback, but then it completely opened my world to how curious octopuses are. And that still fascinates me as a scientist: Why does an animal that's so vulnerable—they have no bones, no teeth, no claws to protect them—want to interact with you? Why does their curiosity outweigh their fear?

One theory, she says, is that their short, solitary lives mean they are "living in the fast lane. They have no parental care, no close social bonds with the opposite sex. They have no one to guide them, so the strategy they've taken is to become curious, rapid learners." In other words, they're interested rather than sociable (I've been to many parties, I tell her, where I've felt the same). "But how they discern between 'this is safe' and 'this isn't safe' when it comes to who they interact with still blows my mind."

And octopuses, she says, are not the only quick study. In 2020, Alex proved that cuttlefish can pass the Stanford marshmallow test (though in their case with shrimp, not marshmallows): a test of delayed gratification wherein a child or animal can choose between an immediate treat or waiting for a better one. "I was expecting them to last a few seconds, like apes and pigeons. But my best subject could wait two and a half minutes. That's massive."

To gain the trust of her subjects, Alex tells me, she hand-feeds and spends time with them. "I turn up two or three times a day and show them they're in a safe space. You need to create a trusting relationship with these individuals for them to be relaxed and exhibit normal, natural behaviors around you. You don't want them to be scared or anxious. Some can be skittish. Others, though, are like, 'Me, me, me!'

Like octopuses, she tells me, cuttlefish are good at making their feelings known, both through the color and, incredibly, the texture of their skin.

"'Back off.' 'I'm angry.' 'I'm feeling timid.' They'll tell you all these things intentionally through color change in the blink of an eye. They use body language and posturing too: They might stand up tall if they're angry. And they can camouflage using their skin's papillae—little bumps like we have on our tongues. They'll extend them to make themselves look rough or relax them to look smooth."

Mimic octopuses, *Thaumoctopus mimicus*, Alex says, are likewise masters of disguise. "They seem to take into account who's watching them when they camouflage themselves, which suggests they're really considering their audience. There's no scientific evidence for this yet, but that's theory of mind."

Some of Alex's cuttlefish, though, have found more unequivocal ways of getting their message across.

"One particular cuttlefish in my lab was really, really grumpy. If I showed up in the afternoon, she would just be waiting patiently, because she knew that's when I'd feed them dinner. But if I arrived earlier in the day, ready to do any experiments, she did *not* want to participate and would aim her siphon up into the air and drench me like a waterfall. And these were all positive-reinforcement experiments, where the animal gets *rewarded* for participating."

"But despite their complexity," Alex says, "people still don't feel comfortable with me giving cuttlefish names in scientific papers, the way that so many chimp and parrot papers do—Alex the parrot, for example. People consider them a 'lesser' animal. Which," she grimaces, "I hate."

Back to octopuses. A 2010 study, Alex tells me, found giant Pacific octopuses like Sedna could recognize human faces and were more likely to approach an experimenter who had previously fed them versus one who had touched them "with a bristly stick." And we're not the only species with whom these legendarily solitary animals form bonds. "A colleague of mine studies Red Sea octopuses, who partner with various types of fish to hunt for prey. Which is also bizarre."

After our call, I look this up. Grouper and coral trout, I read, use cross-species "referential gestures" to signal hidden prey locations to "cooperative hunting partners," including octopuses and giant moray eels. Stranger still, Alex's colleague Eduardo Sampaio has filmed octopuses *punching* these interspecies hunting partners. Perhaps, Sampaio says, it's to sanction the fish for not working hard enough or to keep them from taking the lionfish's share of the spoils (I should think myself lucky then that Sedna only recoiled).

Octopuses also, I discover, use tools. Maybe, I consider, they're using these fish as *living* tools. And perhaps that's what these canny creatures are doing with us in captivity too.

A snippet of Mark Twain pops into my mind. "It is just like man's vanity and impertinence," he said, "to call an animal dumb because it is dumb to our dull perceptions."

If I ever have an octopus, I think, I'll name him Aristotle.

It's mid-December, the season for good news stories, when a splashy headline involving another Twain catches my eye. A group of researchers on a boat in southeast Alaska, it says, has held the world's first-ever twenty-minute conversation with a whale, a humpback named Twain. Whales, unlike octopuses, are vocal animals. We know they have songs. We know these songs change over time and geography through cultural transmission, like human oral storytelling traditions. But still their groans and calls and whistles remain, to us landlubbers, mysterious and out of reach.

In 2018, in the Pacific Northwest, where we used to live, I and thousands of others watched woefully as a wild orca called Tahlequah carried her dead infant on her back for seventeen days, clearly deep in mourning. This gesture seemed to touch people more than whale song ever has. Perhaps that's because whales' sounds are too impenetrable, too elemental, more part of the earth's fabric, like the creak of icebergs or groan of glaciers, than noises with which we're familiar from terrestrial creatures. Their songs are eerie, infrasonic: They are enigmatic strangers from another, older time, inhabiting a world we'll never know.

Certainly, no one has claimed to have *chatted* with a whale before.

I skip the news articles and go straight to the science: a paper titled "Interactive Bioacoustic Playback as a Tool for Detecting and Exploring Nonhuman Intelligence." It's a collaboration between several whale-related organizations and the SETI Institute, standing for Search for Extraterrestrial Intelligence, whose aim is to develop "intelligence filters" in the search for extraterrestrial life.

Listening to whales to listen to aliens?

I visit SETI's website.

"An important assumption," says Dr. Lawrence Doyle, the paper's coauthor, in the institute's press release, "of the search for extraterrestrial intelligence is that extraterrestrials will be interested in making contact and so target human receivers. This important assumption is certainly supported by the behavior of humpback whales."

Instantly, I'm irritated. How could we possibly know that aliens, if they exist, communicate anything like whales, no matter how otherworldly their sounds? Why not porcupines? I fume. Why not rhinos or blobfish or goldendoodles or shoebill storks? Why not *earwigs*?

I had a similar reaction once on a road trip through Roswell, New Mexico, past alien-head lampposts, a store selling "Abduction Jerky," and a sign welcoming travelers to "Spaceport Roswell." We shouldn't care so much, I thought crossly, about listening for aliens when we barely bother listening to the species inhabiting our own planet. If I were an alien, I thought, I'd avoid us altogether.

"Keep going," I growled to the children's disappointment. "Don't stop."

But now I remember Marc Bekoff's "nonjudgmental listening."

I read on.

On an August day in Frederick Sound, Alaska, the paper explains, scientists broadcast playback of a locally recorded humpback "whup" call via an underwater speaker, thereby attracting the attention of a thirtysomething humpback whale named Twain. Whups, sounds like a deep burp combined with a whizz of xylophone, are believed to be "contact calls": a sort of individualized "Here I am!" like red squirrel Sven's, used as social calls between humpbacks of the same and different groups. Twain approached and circled the boat. For the next twenty minutes, they continued to play this "Here I am" call, and taking turns, she replied with whups of her own.

Twain whupped.

There was a pause of several seconds, then the scientists whupped back.

Twain matched, in seconds, the length of each pause before whupping again.

The researchers weren't sure to whom the whup they were playing belonged. It might have been Twain's own, that of a member of her group, or another whale's whup altogether. But in total, they broadcast thirty-eight of them, to which Twain responded thirty-six times.

Afterward, they divided this interaction into three phases: engagement, agitation, and disengagement. During phase 1, Twain approached and engaged, matching her pauses to the scientists' nine out of fourteen times. In phase 2, she emitted three "wheezy or reversed forced surface blows"—considered signs of excitement or agitation— and matched them only twice. In phase 3, she went on her humpback way.

"We believe," lead author Dr. Brenda McCowan said in SETI's press release, "this is the first such communicative exchange between humans and humpback whales in the humpback language."

Hmm, I think. It seems a stretch to call that a conversation.

Sure, there were two senders. There were two recipients. There was a willingness for connection, and there was turn taking. The whups were a medium. But where was the message? Had Mervyn Tugwell sent "Hello" by telegram thirty-eight times, the folks back home might have worried. Are we really conversing if we've no idea exactly what we're saying or what someone's saying back, whose voice we are using to say it, or who the recipient thinks they are replying to?

It seems to me that Twain got interested, then irritated, then left.

"Scientists Share World's First 'Conversation' Between Humans and Whales." Headlines like these say more about our *desire* to understand than about understanding itself. Disappointed, I close the paper and search my phone for a video of a close encounter Indy and I had with another humpback, this one named Kalimba, on a foggy summer Friday.

Brier Island is a tiny East Coast agglomeration of ramshackle fishing huts and dour Victorian manses, straight from the pages of Stephen King. Buoys clank. Seabirds ride, as if for fun, the straits between the island and a smaller lighthouse-laden outcrop.

The weather here is the stuff of folktales. Fog thickens as our boat departs, wrapping us in a gray weighted blanket. Sounds muffle. The ocean's a millpond, the only movement the occasional cormorant flying arrow-straight, low over the water.

It takes an hour to get out to the feeding grounds to which dozens of humpbacks migrate each summer from their Caribbean breeding waters five thousand miles away.

The engines cut.

"Now…listen," says our guide.

Total silence, save for a lap of ocean, a creak of timber, the clink and slap of rope and chain against the hull.

Indy and I stand to starboard, gazing out into a blank, pink nothing.

Then: *whoooooooosh*. The wet, breathy sound of something

enormous coming up for air: the sudden roar of monsoon rains on a metal roof, a rush of steam from an old-fashioned engine.

Whooooooosh.

"They're here," Indy whispers.

More whooshes, now coming from all sides. I hear the sound of spurting water, as if I were blindfolded in a field full of geysers. There are at least a dozen humpbacks, our guide estimates. We are surrounded, but we can't see a thing. Then a crack in the cloud cover. A late sunbeam tears open the fog. And right beneath Indy and me rises a vast, dark shape.

"Is that a shadow?" I frown. "From a cloud?"

"Um…" says Indy, "I don't think so."

We gasp as below us, the shape ascends, grows, pulls focus. A humpback whale, the size of a school bus, is emerging from under the boat. We lean far over the rail, our faces just feet from a great, gray, barnacled back, large enough to lift us out of the water.

"It's Kalimba," says our guide.

Behind it rises a second, smaller shape.

"And her calf: Kalimba 23."

For the next ten minutes, Kalimba and baby hang about beside the boat. We watch, spellbound. These humpbacks are wild. They are not fed or encouraged to come closer in any way. And there are rules: A boat, by law, must cut its engine one hundred meters from any whale it sights, two hundred if it's a mother and calf. This mother, however, has chosen to approach us and to stick around. We look into her eyes. We marvel at her massive fin and flippers, at the golf ball–sized tubercles flecking her beaky rostrum. We hear, up close, the wheeze of her breath. We see every small shift of her forty-ton body.

And then she flukes, deep-diving down for dinner, leaving Baby, who's too young to follow, beside the boat. Kalimba, speculates our guide, expects us to "alloparent"—to babysit—while she dives deeper for fish: A lactating mother whale must eat a ton and a half a day, about the weight of two Majors and an Andvari. She and other mothers, the guide says, have done this frequently before.

Baby seems unconcerned by Mama's departure. She skims the ocean surface alongside our boat. She rolls and blows bubbles and sends up great plumes of spray, playing much like a foal or a puppy.

Then Kalimba returns, and the pair head off, slowly vanishing into the brume.

Of course, I don't know that any of these perceptions were right—that the baby was happy-go-lucky, that Kalimba was really trusting us to calfsit. I can't find any scientific studies to back this up.

But the silent communication that foggy July Friday seemed clear and without agitation.

Reports of Twain's "conversation," in contrast, I decide, have been greatly exaggerated.

Denise Herzing, founder and research director of the Wild Dolphin Project, knows a thing or two about human-cetacean conversation.

"I tried playback once," she tells me, when we meet two days before her project's forty-year gala celebration. "It was 1991, and everyone was all about 'Playback! Playback!' I already had a recorded catalog of dolphins' signature whistles—that's them broadcasting their identity, like 'I'm Denise, I'm Denise!'—so I decided to try it out."

Dolphins' unique signature whistles, developed before each calf turns one, sound a little like R2-D2's chirrupings: closer to birdcalls and too distant from our own vocalizations for anyone other than experts, and perhaps C-3PO, to recognize and interpret. Denise's team attached an underwater speaker to their boat and, with no dolphins in sight, played the signature whistle of a dolphin they'd named Katie.

"Sure enough, Katie's friends showed up thinking she was there.

So I said, 'Let's play *their* whistles,' and we did. One was of a dolphin called Stubby. We knew him well. But apparently we didn't know enough about the etiquette of signature whistles, because he came over to me with a big open mouth and went, 'Raar,' which is aggressive, and then he went over to the speaker and went, 'Raar,' and then he left. And I was like, 'Oh shit.'"

She hasn't repeated this experiment since. "Because perhaps I'm insulting them. And what does it tell you anyway? I learn more by observing them than by playing back their own sounds."

For four decades, Denise, who is warm, wry, and witty, has been communicating with Atlantic spotted dolphins, *Stenella frontalis*, in the balmy waters of the Bahamas.

"We work with friendly wild dolphins who are fairly tolerant of humans. It's a small group of maybe one hundred animals. We know them all.

"It's hard," she says, "to experiment in the wild, because it's very uncontrollable. But I brought out some colleagues who were excited about setting up a device that would bridge the dolphin-human language gap. We thought, let's create a mutual system, agree on a few terms, and play together. And through that play, let's empower the dolphins to ask for things."

They came up, she says, with a series of artificial whistles, each standing for a toy—a scarf, a rope, sargassum seaweed—with which the dolphins wanted to play.

"These sounds weren't already in their repertoire, but they could reproduce them. We then used Irene Pepperberg's method of modeling in the water with our equipment, demonstrating how to ask for these toys from one another while the dolphins were watching. They were really curious about it and learned those new whistles to be able to get the toys."

"That's amazing!" I can't contain myself. "You're adding new words to wild dolphins' vocabulary, *in the dolphins' own tongue.*"

Denise grins. She added herself too, she tells me, in the form of that "I'm Denise" signature whistle.

"We can produce and recognize them all via a translation device called CHAT: Cetacean Hearing and Telemetry, developed for us by a team at Georgia Tech."

Many dolphins who choose to play, Denise continues, are teenagers. "They don't yet have the complexities of adult life to contend with, and they have time on their hands."

"That sounds familiar." I think of my winter-breaking teens, sitting strumming guitars while I'm racing about juggling work, laundry, piglets, dinner.

"So you've got to make sure you're not boring, because they've got interesting lives and might otherwise be like: 'OK, I'm going to go harass a turtle instead.' It's about them *wanting* to communicate and maybe getting from you something they can't from their natural environment. They want the proverbial cookie."

Or the proverbial pill bottle, I add, for an octopus.

But Denise's team, she says, isn't only creating a "bridging" version of Dolphinese; it's also studying and attempting to translate dolphins' own full language of whistles, clicks, and burst pulses.

"Like everybody else right now in animal behavior apparently, we're using AI and machine learning to look for patterns and language structure, to try to 'crack the code.'"

I've noticed that trend too, I tell her. I've lost count of the dozens of clickbait headlines I've come across, claiming we're only a whisker away from full two-way communication with marine mammals. And in June 2023, just five days before he died, Roger Payne published an impassioned *Time* article in which he advocated for the work of Project CETI, the Cetacean Translation Initiative, which is using AI to try to decipher what Payne called "WhaleSpeak."

It's naturally very tempting, I tell Denise, to believe there could be a translation device that will transform cryptic communications from the deep into something the everyday, interested human can understand. But is it true? *Can* AI bridge the aquatic divide between us and these linguistically far-distant wild-water creatures?

She smiles drily. "Machine learning excels at mining massive sets of data. It helps you extract and label stuff from your recordings that would otherwise take you weeks, months, or even years. It can find sequences and repeated patterns: things you'd expect to be underlying if a species has a language. And boy, the field sure is changing quickly."

But there are, she warns, many overblown claims being tossed about.

"Even if we found some cool sequences that would suggest language, how do you test this? If you were doing this with another group of humans—say, an Amazonian tribe—you'd want to go in and say that sentence and see what happens, what the meaning is. But that *interpretation* component is what AI's missing. If you have enough data, you could speculate: 'If dolphins vocalize pattern A, then thing B mostly happens.' But it's still a big 'if.' AI's not some magical thing where you throw in sounds and out comes a language."

And even human language translators, I add, could use some tweaking. This morning, idly browsing ponies for sale online, I found two advertisements autotranslated from Icelandic.

For sale a beautiful yellow starry horse glazed on both, read the first. *Born this summer. Very well said. Became a bandit and boring. For more information: in the sky.*

She flies up on a cart, said the second, *and no problem ironing. She is not a little fax puppet. She is a slave inside the palace (see video).*

"Unlike a human," says Denise, "AI can't consider individual context. What's the situation? What's going on between these individuals? What's their relationship, their common ground, their body language? For example, we once brought with us our captain's baby.

The dolphins had never seen little people before: They were fascinated and brought their own calf over. That's valuable communication right there. Half of speaking, even with chatty dolphins, isn't about vocalization at all."

Like that old joke, I tell her: Tie an Italian man's hands together. He'll be speechless.

Over the next week, between wrapping gifts and making three kinds of stuffing, I trawl through more pros and cons of AI. How it could enable us to follow an individual animal, à la *The Truman Show*, through its entire life, mark the exact moment a baby bird first sings its learned songs or the sound it makes every single time it spots a human. How it already helps connect us to the animals in our environment through identification apps like Merlin for birdsong and Happywhale for the individual fluke, or tail, patterns on whales, allowing scientists to track their paths across oceans.

But I consider how science doesn't yet know what, if any, the smaller units of speech are within whalesong or Dolphinese—how if we were to analyze them with the wrong segmentation, they would become meaningless. (This reminds me of my favorite Gary Larson cartoon, in which a team of scientists is studying two dolphins in a pool. "Matthews," reads the caption, "we're getting another of those strange 'aw blah es span yol' sounds.")

And how, I wonder, will AI handle the ever-evolving songs of our oceans' whales or the local dialects of bats or prairie dogs?

This brings me back to something Irene Pepperberg, who trained Alex the African gray, told me. "AI is a probabilistic system," she said. "It can predict, when you're typing an email, that if your first three words are this, then the fourth will likely be that. Likewise, it can analyze animal communication and say, 'After you hear these three notes, this fourth note is likely to happen.' But it can't ascribe meaning where no human has yet discovered it, and that takes incredibly careful, in-person research."

Besides, she added, say you do crack the code. "What do you do with the bad players who will use it for poaching and illegal hunting and things like that?"

Next, I learn some astonishing things science would never have discovered without human ears and eyes firmly on the ground.

How no one knew elephants spoke in infrasound until 1984, when zoologist Katy Payne, wife of Roger, first felt their communications as a throbbing in the air and pulsation in the ears, "sort of the feeling you get," she said, "when your windows are rolled down in your car." Under the right atmospheric conditions, subsequent studies found, these conversations can travel six miles.

How in 2012, biologist Camila Ferrara discovered, with just her ears and a microphone, that South American river turtles, *Podocnemis expansa*, who until then were believed silent, talk to one another from within their eggs to coordinate a moment of mass hatching while their mother, famously thought to abandon them, waits in the water, calling them to safety.

How researchers noticed happy dogs tend to wag their tails more to the right and anxious dogs to the left and, watching dogs watch dogs wagging, found that dogs notice too.

After a morning spent watching dogs' tails wagging at our local Christmas farmers' market, I even attempt a little AI experiment of my own.

Winnie, home for the holidays and unhappy about it, has been hanging about upstairs at the farmhouse. I consider draping the beds in plastic sheeting, but opt instead to upload MeowTalk, "the world's #1 science-backed and AI-powered cat translator app," which cat-egorizes meows into eleven "general intents" ranging from *happy* to *angry* to *give me attention*. Perhaps this way, I consider, I will find out what's the matter.

Cats themselves are excellent listeners. A 2022 study shows that domestic felines can distinguish between human speech addressed to them versus to other humans. Meanwhile, their own voices, like dogs', are more often directed at people than at one another. But context is key. Even purring can mean a host of things, from contentment to irritation to imminent death, and one 2020 study suggests we owners "are not good at extracting precise information" when it comes to cats' vocalizations.

To brush up on a range of them, I visit the website of linguist Susanne Schötz's Meowsic cat communication project. Here, common cat sounds are divided into three types: sounds produced with mouth closed (purrs, chirrups), sounds produced with an opening-closing mouth (yowls, meows), and sounds produced with an open, tense mouth (growls and hisses). At the bottom of the page, a button reads, "Cat Vocalization Types in Other Languages." I click on it and, disappointingly, get the same information in German.

Winnie, however, has always been the strong, silent type. With persistence on my part and tolerance on hers, I elicit two purrs and even some halfhearted moments of play, but day after day, she gives MeowTalk nothing to work with.

"Come on, Winnie. Say something. Anything, Winnie. Winnie?"

She stares right through me, a squash-faced sphinx. The SETI people, I think, should have a go with her.

Downstairs, the vacationing teens are watching *Notting Hill*.

"'You say it best,'" I sing along, "'when you say nothing at all.'"

Winnie yawns and remains noncommittal.

Finally, I read how two weeks before Christmas 1973, when the Carpenters' "Top of the World" topped the Billboard chart, a neat, bespectacled Dutchman took to a podium to deliver a lecture he'd never dreamed he'd be giving. His name was Nikolaas "Niko" Tinbergen, a biologist with a passion for birds, who had, to his tremendous surprise, just received the Nobel Prize for Physiology or Medicine along with colleagues Konrad Lorenz and Karl von Frisch. It was the first time animal scientists had ever won the award.

"Many of us ourselves have been surprised," he said, "at the unconventional decision of the Nobel Foundation to award the year's prize for physiology and medicine to three men who have until recently been regarded as 'mere animal watchers.'"

He went on to extol the virtues of "open-minded observation" and "watching and wondering." "This basic scientific method," he said, "still too often looked down on by those blinded by the glamour of apparatus, by the prestige of tests."

It's tempting to believe in promises of a quick fix, that we'll be chatting seamlessly with cetaceans before next Christmas rolls around. But behind the glamour of the headlines, AI is just another bit of clever, useful kit. I remember how, in *King Solomon's Ring*, ethologist Lorenz, Tinbergen's Nobel corecipient, quoted Shakespeare to describe how most of his learning had derived not even from books, but simply from watching nature. The words he chose are from *As You Like It*:

> And this our life, exempt from public haunt,
> Finds tongues in trees, books in the running
>   brooks,
> Sermons in stones, and good in everything.
> I would not change it.

The next day, the farm drifts in a dreamy haze. A string of fairy lights tints the mist outside the kitchen door. Black-capped chickadees ornament a thornbush. I step outside, and the world sounds weird:

some things muffled, others amplified. I hear giggles and *whees* from the starlings in the oak, an eerily clear *moo* from a cow across the river. Siegfried's calls from the wood are poltergeistish.

My phone rings. I jump. A number I don't recognize.

"Hello?"

It's my early Christmas present: a phone call from the legendary ethologist Temple Grandin to share her thoughts on listening to the silenter animals. She is sitting aboard a United Airlines plane from Phoenix home to Denver and has had a run-in with the limitations of technology. Her first plane—"some old ancient antique," she says—couldn't get the door hydraulics to operate, forcing her to change airlines and terminals to try to get home, "a great big pain in the butt."

"I am an extreme visual thinker," she says as we get down to business. "When I first started out, designing humane cattle chutes for slaughterhouses, I just looked at what the cattle were looking at, at what they were seeing, and from there came up with my plan. To understand animals, *we need to get away from verbal language.*"

That's why AI, she says, is fallible, especially given its tendency to make things up.

"There's that case of the lawyer who used a ChatGPT-created judgment in court—only to find it didn't really exist. We will never," Temple says firmly, "find animal language just using machines. Animals definitely have emotions. They definitely try to communicate with us. But we must listen, in person, to the sensory-based world, not the word-based world. I can't stress that enough. It's the sensory-based world where the talking really lies."

I hear the pilot's announcement, tinny, inscrutable, over the airplane's speaker.

"They're closing the doors," she says, successfully this time.

I wish her a pleasant flight, as Temple Grandin, who sees the world as an animal sees it, takes to the skies in a clever, intuitive, man-made object that still needs piloting by a person.

It's Boxing Day and a rare sunny afternoon when Gal steps outside to take a conference call, letting the piglets out to stretch their six-inch legs.

He is wearing an old Sex Pistols T-shirt, a pair of basketball shorts, a huge winter coat, and rubber boots. With one hand, he clasps his cell phone to his ear. In the other, he holds a half-eaten peppermint candy cane.

Passing the kitchen window, I stop to relish the scene. I think of other lawyers in their three-piece suits in New York, Beijing, Birmingham. No one would imagine what this serious, fierce-sounding specimen is really doing.

But Constance knows exactly what he *should* be doing, and that is petting her.

First, she grunts.

He doesn't hear her.

Next, she tries for attention by nudging his knee.

Nothing happens.

She thinks for a second, then sticks her moist, mobile snout determinedly down one of his wellies.

Still nothing.

She takes a few steps back, assesses the situation, and trots at him, giving him a gentle headbutt in the back of one leg.

Wrapped up in his call, he shifts his weight and turns away.

This is the last straw. With great purpose, Constance returns to his front, positions herself side-on to the rude and uncommunicative wellies, and determinedly capsizes on top of them.

Pet. Me.

Gal staggers, finds he can't move either foot, looks down, and, to his apparent astonishment, sees a pig. I watch him mute his phone, crouch down, and give Constance the scratch she was after.

She closes her eyes. She heaves a big sigh. He continues his negotiations with candy cane in hand and a fifty-pound pig asleep on his feet.

"Silent" creatures, I think as I surreptitiously take a photo, are not silent at all. Science continues to find new voices—though like turtles' and elephants', they've been here, talking quietly all along. There must be many more still to be heard. But there are no shortcuts for this, no quick fixes. Scientists like Temple and Denise and Ahmad agree that the key to hearing the unhearable is still Niko Tinbergen's in-person, open-minded listening. Open-minded enough to find commonality with the seemingly most solitary, phylogenetically remotest of creatures, to think of vibrations as a voice, to watch for body language in bodies utterly unlike our own, and sense connection and companionship when others might maintain it isn't there.

Even a five-month-old piglet knows the value of "watching and wondering," and that actions, on occasion, speak louder than words.

Treading the Boards and Lessons from Apes

How Much Listening Is Too Much?

> "For it is the destiny of man that he should seek to take upon himself the burden of understanding, and to move in the comprehension of his works and the consciousness of his crimes."
>
> —JOHN STEWART COLLIS,
> *The Worm Forgives the Plough*

JANUARY

I contain multitudes.

Whitman's words spring instantly to mind when I catch my first glimpse of Loulis and Sue Ellen amid the raised wire walkways of the Fauna Foundation in the brumous depths of the Quebec countryside. Sue Ellen, a fifty-seven-year-old chimpanzee, sits eating a frozen purple smoothie, observing construction workers come and go from the building (she especially likes watching men, primatologist Dr. Mary Lee Jensvold tells me). Loulis, age forty-four, lingers farther away, just within sight.

"Want to come here, Loulis?" Mary Lee calls up to him. "Say hi?"

Loulis makes his way over, a handful of nuts clutched in one fist. Mary Lee performs a chimp "pant greeting," a series of short, sharp, shallow inhales, and signs a salute-like "Hello" in American Sign Language, or ASL. Too shy to try the chimp version, I sign too.

Loulis sits down and screws his eyes shut.

"That's his equivalent of putting his fingers in his ears," Mary Lee tells me, "saying, 'La-la-la—I can't hear you.' Unlike Sue Ellen, Loulis doesn't like the workmen. He's had a hard day."

Loulis opens his eyes and signs hello. Though he might not know it, he is unique: the only animal on Earth known to have been taught human language *by another animal*. His teacher was his adoptive mother, Washoe, a chimp used in one of the twentieth century's famous ape language experiments that sought to establish whether and to what extent primates could acquire and use human language.

Loulis knows and uses around sixty ASL signs. His older sister, Tatu, who's currently out of sight though is never, says Mary Lee, too far away, knows and signs about 250. Sue Ellen, meanwhile, is not an ASL chimp, but a survivor of biomedical research.

"For fifteen years, she performed for a circus," Mary Lee says as Sue Ellen peers down genially, "then they sent her to a lab where in just one year, she endured twenty-nine liver biopsies. She was infected with HIV when scientists were seeking a cure for AIDS. She had more liver biopsies and rectal and lymph node ones. In total, she was there for another fifteen years."

It might seem, compared to this, that the apes involved in language research fared far better. Most were cared for, played with, offered physical and mental enrichment and plenty of attention. They appeared in *Time* and on *Sesame Street*. For a while, they were darlings of the press.

Look closer, though, and a sadder story emerges, beginning in the 1950s with research psychologists Catherine and Keith Hayes and a baby chimp named Viki. Great apes, our genetically closest relatives, seemed to the Hayeses the obvious choice to find out if any animal could learn to speak Human. Human babies, they reasoned, acquire language through being raised in a family home, and they set about teaching Viki the same way.

It's painful to watch the resultant footage, which runs like a social guidance film: a prim middle-class couple, sitting with their neatly dressed toddler at the breakfast table. "Papa, Papa," says Keith Hayes. Viki copies by smacking her lips, shaping them with her fingers to form the words *Mama* and *Cup*. She hasn't the ability to fully do what's being asked of her. Apes' vocal mechanisms, researchers afterward realized, are not similar enough to humans'. Viki seems belittled in her pink tunic: a parody of a human person, like our local zoo's chimpanzee tea parties that I detested as a child.

Five years later, when Viki will die of viral meningitis, she will have managed a spoken vocabulary of seven words.

No one repeated this experiment. Instead, across America, a raft of researchers including psychologists Herbert Terrace, David Premack, and Penny Patterson and primatologist Sue Savage-Rumbaugh set out to teach the use of lexigram symbol boards and ASL to chimps, gorillas, and bonobos. They sought two things: symbolic reference (using an arbitrary sign to mean something, the way we humans do with words) and basic syntax (stringing those words together in a meaningful, sentence-like order).

The most successful of these ventures was Project Washoe, led first by Drs. Trixie and Allan Gardner and later by their students, Drs. Roger and Deborah Fouts, who taught ASL to a group of chimps including Washoe—who died, aged forty-two, in 2007—Loulis, and Tatu. Mary Lee was both the Gardners' and the Foutses' graduate student, eventually succeeding Roger as director of the Chimpanzee and Human Communication Institute at Central Washington University. She is also one of the few people on Earth to have been interviewed for a job by a gorilla.

Koko (1971–2018) was a Western lowland gorilla to whom Penny

Patterson claimed to have taught over one thousand ASL signs. Tasked in 1985 with choosing her own assistant, Koko entertained then-student Mary Lee in her trailer home in the Californian grounds of the Gorilla Foundation.

"I had brought a couple of gifts for her," Mary Lee tells me, "like a handkerchief and a hat. I first met her through the sliding glass door of her trailer. She invited me in, so in I went. Inside was a chain-link fence. Koko was on one side, and Patterson and I were on the other."

Mary Lee handed Koko the hat.

"She called it a 'toilet hat.' I don't know why—maybe she didn't like the way it smelled. Then she went away and filled her mouth up with something. I'd never been around a captive ape before, but I knew she was up to mischief and thought she was probably going to spit it at me. So I told her she was tricky or devilish—something like that. And she seemed so shocked that I'd insulted her."

Mary Lee did not get the job.

"Nevertheless, it was the most transformative experience. I stood waiting for my flight home, thinking. I've grown up in this culture that sees humans as fundamentally different from nonhumans, but I've just had a *conversation* with a nonhuman. So where am I going to draw that line now? Does it go between apes and monkeys, or between primates and nonprimates? Then I realized. That line didn't need to be drawn at all. Evolution doesn't have sharp boundaries. It's a continuity."

A *continuity*. I love that, I tell Mary Lee. Because as Darwin intimated and as I've found so far in animals from fruit flies to octopuses, there's no "us and them": rather, infinite varieties of "us."

Instead, Mary Lee went to work with the Foutses and Washoe, investigating chimps' imaginary play and, fittingly, how to overcome misunderstandings in human-ape conversations.

Over the years, Project Washoe yielded incredible results. The chimps acquired and retained hundreds of signs and used them appropriately in conversation with multiple humans. They astonished their human caretakers by routinely outperforming their expectations. They produced, like Alex the parrot, novel phrases. The first time Washoe saw a swan, she called it a "water bird." Tatu coined "bird meat" for Thanksgiving turkey and asked, "Time sweet tree?" referring to edible Christmas tree decorations. Another chimp, Lucy, signed "cry hurt food" when she first tasted a radish.

The chimps signed to each other. They signed while alone, as if thinking out loud. Washoe taught Loulis to sign. I watch clips of them conversing with humans. Unlike Alex the parrot, whose days comprised hours of tests, repetition, and training, they discuss everyday things: games, food, visitors, pictures in a magazine. They could talk about the past, things that happened both yesterday and years ago.

"When Tatu moved to Fauna," Mary Lee says, "there was a chimp here named Yoko who was very ill. The day after Yoko died, Tatu pointed to Yoko's room and signed to me, 'Hurt. Hurt in there.' We hadn't taught her a sign for death, so she was sharing information the best way she could." And Tatu, she says, still sometimes mentions Dar, a chimp who died in 2012.

"You can ask the chimps a question in spoken English," she continues, "and they'll respond in ASL. If there's a misunderstanding during your conversation, they'll adjust and persist, and use different words to get their point across. They're very attuned to their partner. That's sophisticated perspective taking."

As a graduate student, Mary Lee remembers, she often worked weekends alone. "At the end of the shift, I'd think, 'I've been having all these interactions without ever opening my mouth, and *they're not humans.*' It never fails to amaze me."

But as understanding of the chimps' abilities grew, so too did the uneasy feeling among Project Washoe researchers that they should not be engaged in this research at all.

"Suddenly I realized: Wow, I spend all day having conversations with them and caring for them and keeping them in cages, and they have no hope of ever getting out. The more you appreciate what thinking beings they are, the depth of their perception of our world, the more you also understand the depth of their suffering. I felt it especially after becoming a parent. I could give my child all these opportunities, but the chimps, I couldn't. I'd get back from a great trip and think, 'Oh, you're still here, and you're still here, and you're still here…'"

For Mary Lee, the conclusion was clear: Project Washoe should never be repeated.

"Even now, we still have so much video footage and handwritten data. We can use that to tell people what it's like to be a chimp, because they're telling us their experience. But it's unique. It's a one-off. Because we must *never* do these experiments again. These two, Loulis and Tatu, are it."

They are also the lucky ones, spending their "retirement" in relative comfort at Fauna. Others fared less well. "Lucy, they sent back to the wild in Africa. She quit signing. She was really depressed, and eventually she was poached. It's heartbreaking." Another chimp, Booee, was sent for medical testing, kept caged, and infected with hepatitis C, signing "Key" to captors who could not understand him and wouldn't have let him out if they had.

One good thing to come of it all, Mary Lee says, is that in 2015, the U.S. National Institutes of Health discontinued medical testing on chimpanzees. "But it wasn't us who paid the price for our learning." Testing on smaller primates, meanwhile, still goes on.

These days at Fauna, caregivers communicate not only in ASL, but also in Chimp.

"Positive communication with humans," says Mary Lee, "can help treat these survivors' trauma. They still know we've got the keys and that we're never going to let them out. But I think a relationship with humans is a pathway to making their day a bit better. A lot of human behaviors, though, can be unintentionally threatening to chimps. Our bipedal stance, for example. And a toothy smile to us is a fear grimace to them. So we teach our caregivers chimpanzee play behaviors, grooming behaviors, to do lip smacking and mouth sounds. And when that relationship is working, it can bring such light to the chimps. It can really lead to healing."

Up above us, Loulis comes closer.

"What are you doing?" Mary Lee asks, both in voice and sign.

In reply, he signs "Chase."

"You want me to chase you?"

"Chase," Loulis signs again and races away.

The afternoon light is dwindling. A quiet's setting in. Inside, the chimps browse magazines, doze, and play with enrichment activities. There's an air of old folks' home: comfortable, functional, a little down-at-heel. The foundation will close, Mary Lee says, when there are no chimps left.

"Are they leading full lives, do you think?" I ask.

"The problem is that the possibilities are so limited. It's very hard to see them like this."

Sue Ellen follows Loulis inside for the evening.

I watch them go, these individuals with such complex lives, such resilience and forbearance, who learned to exist in our human world in the roles we assigned them. ASL allows us glimpses of their lived experience: rich, multilayered, feeling just as keenly, sensing just as acutely, understanding just as readily as ourselves, but kept captive their whole lives in ways we could never tolerate. I'm amazed they are willing to communicate with us at all.

I am large, I think of Sue Ellen as she vanishes into the gloaming. *I contain multitudes.*

We head down Fauna's driveway past a five-foot-square wire cage.

"Was that used for transport?" I ask.

Mary Lee shakes her head. "Chimps lived in those in the biomedical labs. Suspended from the ceiling in rows. No floor—just wire underneath. They tranquilized them to hose the cages down."

"Oh." We walk on in silence.

"I went back to see Koko near the end of her life." Mary Lee turns to me as we reach my car. "Thirty years had gone by, and she'd been in that same trailer the whole time. She'd met a lot of famous people. She was known all round the world. And yet there she still was. It was really, really sad."

Primatologist Cat Hobaiter's average day at the office is the stuff animal-loving dreams are made of. Young, funny, raven-haired, Cat spends half her year at the University of St. Andrews, Scotland, and the other half embedded in groups of wild chimpanzees in the Budongo and Bugoma Forests of Uganda and the Moyen-Bafing National Park in Guinea, or with mountain gorillas in Uganda's Bwindi Impenetrable Forest, studying the subtleties of their rich gestural communication.

"On an ordinary morning," she says when we chat on Zoom, "when they first come down from the nest, I'll just take it easy, give them a bit of space."

"Like with teenagers?" I ask from experience.

"Exactly. I spend the first couple of hours hanging out with them, settling in, finding out what we're going to do today. That easing-in process is important."

Cat's research is rooted not in the experiments of the Gardners and the Foutses, but in the fieldwork of Jane Goodall, who famously proved that the wild chimps of Gombe, Tanzania, could both make and use tools.

But Cat's approach differs from her famous forebear in one

important respect: "My goal," she says, "is to be really, really boring. I try not to influence their decisions. I'll often just plonk myself down against a tree to listen and pretend to be fascinated by those leaves up there or some silly such thing. Inside, I might be doing a happy dance at what I'm "hearing" by observing, but outside, I'm channeling my bored fourteen-year-old self who just doesn't care."

Very young chimps who grow up around her, she says, will test her resolve. "Aged three or four, they'll shake a branch to see what I'll do, and there was one little girl who'd drop sticks on my head. I know if I react, it becomes the best game ever, so I just have to sit there like an idiot. But usually that's just a phase the younger chimps go through, and then they get older and discover boys or politics or the grown-up world of being a chimp. You're just not that interesting anymore—and that's my goal."

This proves trickier, she tells me, with the mountain gorillas. "They're so wonderfully relaxed, and I have the distinct impression I'm used as entertainment for the kids. The mums will walk along, literally shrug off Baby right in front of me, then go climb a tree to feed."

Like Kalimba, I remember, dropping off Kalimba 23 for playtime beside our boat.

"And I'm just sitting there on the ground with a baby. And Baby is like, 'Oh, you're a human! You're the best thing ever!' But even then, my goal is to be a vaguely neutral presence."

Cat's work focuses on subtlety: both in everyday ape gestures and in gestural dialect differences between communities.

"In human language, you could use exactly the same words—like 'OK bye' or 'See you later'—and it could mean something super insulting or something really friendly or anything in between. You could even use the same intonation and tone, and its meaning would depend on what's just happened between you."

But with much nonhuman species' communication, she tells me, science has focused on "universal meaning." What's the general meaning, we ask, across all individuals and situations? We then gather a big dataset and, Cat says, "lump it all together."

"That's exactly the opposite of what we instinctively do as humans, which is to interpret communication in situ—find the occasion meaning. So at first, with the apes, we'd say, 'I have four hundred uses of this gesture. What does it mean?' Now we look at it more like 'What did Frank mean on Tuesday at 2:15 p.m. when he used it with Fred after they'd just had a big argument?'"

From this, the Great Ape Dictionary was born: a visual, gestural dictionary to help humans understand what apes, in various contexts, are saying. "The dictionary," Cat says, "is being used by researchers, psychologists, biologists, and ecologists, but also by artists, dancers, and video-game designers as inspiration for an array of creative pursuits.

"For every gesture we record now, we describe about forty different things: Who gave it, who is it to, where were they, when did they use it, why did they use it, how fast was it, what body part did they use. All this incredible richness. We tend to be trained as scientists to think of variation as noise in the data, a problem to get rid of, which makes animal communication look simpler and less flexible than it is. But by just looking at what's generalizable—'What does a human do?' 'What does a chimp do?'—we're getting rid of the most interesting bits of their world."

There are, however, some universals. In her 2023 study, she tells me, she found that everyday people are excellent at discerning the meaning of chimps' and bonobos' gestures.

"I've spent decades becoming an expert at decoding chimpanzee gestural communication." Cat grins. "And it turns out people with no experience can do it quite well. On one hand, that's supercool science. On the other hand, it's a bit irritating."[†]

[†]   Though the study has wrapped up, Cat says, her team has left its quiz for interested humans to try on the Great Ape Dictionary's website (https://www.greatapedictionary.ac.uk). We

Chimps and gorillas, she explains, share 85 to 90 percent of their gestural system. "We humans are chimps' closest relatives, so if these guys have so much of it in common, surely we humans must have access to it also. Our question was: Has language replaced it? We made the case that it hasn't, that we clearly, on some instinctual level, still understand it.

"The finger of shame," she tells me, wagging her finger, "works on any animal. And my friend took a video of a male chimp playing airplane with his baby. It went totally viral. People can see themselves in it, which is really powerful for conservation and for the welfare of captive individuals too."

"But what about that old Nagelian 'What-is-it-like-to-be-a-bat?' dilemma?" I ask.

"Right. Even if you could speak perfect lion, would you know what it's like to be a lion? That's why for me it's important to just spend time with them, embedded in their world, to understand what it means to live in a rainforest, to be a chimp, to be a gorilla. The point of communication is not the system. It's not the language. None of that matters. The point is to share ideas."

It thrills me, I share, that she does so by listening to apes in ape habitats and not by teaching Human to apes in ours.

These days, though, Cat tells me, the Great Ape Dictionary has assumed a less heartening function. "We now call it a 'data ark,' because we're dealing with endangered species, and many of these groups won't exist in the future. So I think of it as cultural preservation, like the people impelled to record dying human languages. When we talk about great-ape conservation, we often use numbers: 'There are five thousand left' or 'There are ten thousand left.' But if you lose one single community of one hundred chimps, that's an entire distinct language and culture you can never get back."

---

have a go later, and sure enough, both Indy and I identify ten of the fourteen gestures "like," the test congratulates us, "other great apes."

Back out in the field, Cat says, she knows she can never be 100 percent boring.

"My presence will inevitably shape some decisions the apes make. For example, I know I'm the chimps' secret weapon if we go off boundary patrolling, because they know the neighbors, who are unhabituated to humans, will be terrified if I'm there. So they'll deliberately hang back and wait for me to keep up with them. I'm fully aware," she laughs, "that I'm being used."

In May 1980, at the New York Academy of Sciences, esteemed semiotician Thomas Sebeok held one of science's strangest conferences. To it, he invited circus performers, the Amazing Randi, a magician turned paranormal investigator famous for debunking Uri Geller's fork-bending mentalism, and ape-language researchers. Sebeok called it a "celebration of deception in all its varieties." Randi demonstrated the principles of confirmation bias. Magic tricks, he explained, work by diverting the audience's attention from what's really happening. Ape-language researchers, claimed Sebeok, were doing the same. They were cuing the apes with "ostensive signals." They were embellishing their findings, overinterpreting their subjects' signs. They could be divided, he concluded, into three groups: self-deceivers, fraudsters, and failures. The majority, he said, were the former.

The fallout tore the burgeoning field apart. Funding plummeted. Researchers turned against one another. The Rumbaughs, Mary Lee Jensvold told me, "threw everyone else under the bus." "It's Kissinger's argument," said Irene Pepperberg, who attended the conference. "Why are academic fights so brutal? Because the pie is so small."

The press picked up the story. Some were reminded of the Nature Fakers wars at the turn of the twentieth century, when naturalists, such as Ernest Thompson Seton and William J. Long, insisted animals demonstrated individuality, while others, like John Burroughs, maintained this was wishful thinking.

The field contracted. Studies shut down. Sebeok's damage was done, largely inspired by the findings of one of his "failures": Herb Terrace, ape-language naysayer, and his infamous Project Nim.

Herb waits patiently for me to get my AirPods working and grasp my pen the right way up. It's January 17, and a horrifying climactic quickstep (rain-freeze-thaw-snow-rain-thaw-freeze) has turned the horses' corral into one of the only unreceding glaciers in North America. Now I'm late for our Zoom call, I've forgotten my questions, there's ice up my trousers and hay in my hair.

"Sorry," I pant, "technical issues." I can't tell him that really, armed with crowbar and sledgehammer, I've been outside attacking the ice and swearing like a sailor. "Right. Ready. Sorry again."

Herb Terrace, eighty-seven years old, self-possessed and dapper in a black turtleneck, gives me a thin, tolerant smile. Once a pupil of the mighty B. F. Skinner, he set out fifty years ago, he says, to prove apes *could* be taught to use human language. He acquired a young chimpanzee, teasingly named him Nim Chimpsky after Skinner's archrival, and set about teaching Nim ASL at Columbia University.

But what he found, he tells me, was twofold.

"The more I looked into it, the more I thought the other researchers' methodology in this field was weak. I looked at the Gardners' work with Washoe and at Penny Patterson's with Koko. There was no video confirmation of what these people were claiming."

And second: "For a while, I was sure I'd succeeded in teaching my chimp language. But then I discovered that just about everything he signed was prompted by the teacher about a quarter second *before* he made the sign. And that threw out any thought of spontaneity on Nim's part—because in fact it wasn't language. Nim had simply learned to be a beggar: a very good one."

Herb had, he says, already submitted a paper to the journal *Science*, claiming apes could learn grammar.

"So in 1979, I retracted it and explained what was really happening—that unfortunately chimps have no interest in language or learning the names of things. If a chimp wants something, he'll just grab it, and if you say, 'No, you have to sign,' then OK, he'll make the sign, but that's the end of the story."

Since then, Herb's beliefs have remained unshakable.

All humans can name things. Animals can't.

All humans take turns as speaker and listener. Animals don't.

All humans can create sentences. No animal can.

Humans use declaratives, animals only imperatives.

Animal signals are innate, not learned, and they rarely possess more than a dozen.

All the chimps who learned ASL, he says, were either rote-learning or being unconsciously cued, and they only gave a conditioned response in order to receive a reward.

"The moral of the ape language experiments," he concludes, "is that it's futile to teach a chimpanzee to produce sentences if it can't even learn to use words."

"I see." I blink, bewildered. Just last week, I was watching, in real life, a chimp sign words. Yesterday, I watched and rewatched videos of Tatu and Washoe and Loulis contradicting almost everything this formidable professor just said.

"Look," says Herb. "There's a difference between *production* of language and *comprehension* of language. You can say to a dog, 'ball,' 'doll,' 'book,' and the dog will go out and fetch those things. Some dogs do it very well. But animals are not hearing words. They're hearing noises. And when they bring the corresponding item, they're given a reward. So there's no word production."

I think of Alex's "chalk" and "banerry;" Tatu's "bird meat" and

Lucy's "cry hurt food." I think of dolphins' signature whistles and Con's prairie dogs' "blue" and "yellow;" how primatologist Sue Savage-Rumbaugh recorded Kanzi the bonobo describe Fanta as "Coke that is orange."

"Oh." I show limited signs of word production myself.

Herb himself, I know, has his critics, who believe his science was at fault, not Nim's ability. That his program was not well designed or even humane, that Nim had dozens of trainers instead of a stable two or three, few of whom were proficient in sign language. That Herb's methods were more interrogation than conversation. That he used extrinsic, nonreferential rewards, which made the association between the sign and what it stood for less relevant for Nim.

"Were you disappointed?" I ask. "With your results?"

"Of *course*. I thought I'd be famous for having a conversation with a chimpanzee. But in science, you've got to call it as it is. If there were no treats, there would be no communication. The story of Dr. Dolittle is very compelling. Everybody wants to talk to an animal. Humans are empathic, more empathic, in fact, than any other creature. So it's easy to anthropomorphize and call what an animal is doing "language." But that's all it is: the result of our empathy."

In 1977, Herb killed Project Nim. He sent Nim back to the facility that bred him, who sold him on to a pharmaceutical lab for animal testing. Following public outcry, Nim was purchased by a sanctuary, where he exhibited manifold, violent symptoms of PTSD. Elizabeth Hess's 2008 book, *Nim Chimpsky: The Chimp Who Would Be Human*, relates how, when visited by an old "friend," chimp researcher Bob Ingersoll, Nim signed, "Bob, out, key." Nim died there of a heart attack, middle-aged at twenty-six, in March 2000.

Outside, temperatures are dropping. I need to get back to my crowbar and sledgehammer. And this conversation has made me uncomfortable; it reminds me how far apart you can be in conversation with a fellow human being, in perspective, priorities, interpretation. Because Herb's right. My human empathy *does* lie with Nim, not because I think he couldn't speak, but because I believe he could.

"Do you have any regrets?" I ask, heart in my throat. "About Nim himself?"

I hold my breath.

"No," Herb replies. "The only regret is that I wish there was some way I could have succeeded. But chimps just don't have the ability or the motivation."

"*Scritch*es."

"Scritch*es*."

"*Scritches*."

After three tries, the word is starting to sound weird, though not as weird as the recorded sound of my voice saying it from deep within a little plastic button. *Scritchess. Scritchies. Screeches.* I sound alternately like a member of the royal family, a 1950s television presenter, a nutcase.

I press Record for the fourth time. "Scritches," I declare and leave it at that. Puff, I'm certain, will know what I mean.

It's one of the darkest, dreariest days of late January, but my day has been brightened by the arrival in the mail of a set of FluentPet Augmented Interspecies Communication, or AIC, push buttons, through which pets can allegedly be taught to communicate their thoughts, feelings, and desires to their human owners.

The idea isn't new. Denise's dolphins, Premack's chimps, and Sue Savage-Rumbaugh's bonobos have all learned to communicate through some version of it. A study I discover on horses finds they can be taught to touch an arbitrary symbol to indicate whether they would like to wear a blanket. One of the first things I ever did with poor, worried

Andvari was install a "go" button: teaching him to press his nose gently to my hand if he was willing to be touched or petted. Major, simply by watching me teach it, learned it too.

But the boards are more famous these days for their use by celebrity canines like Stella the Catahoula and Bunny the sheepadoodle. On Instagram, I watch Bunny look into the mirror, then press buttons that say "WHO THIS." I see her press "OUCH STRANGER PAW," to describe a foxtail stuck between her toes. And, my favorite, I watch her watching a seagull outside the window. "BELLY GO BIRD BELLY," she communicates. "Do you wanna eat the bird?" Alex Devine, Bunny's owner, laughs.

Now it's Puff and Dolly's turn—though we're starting small.

I load the first chunky plastic button, which resembles a 1970s game-show buzzer, with a word containing, as for both Alex the parrot and the Project Washoe chimps, an extrinsic reward.

"Scritches."

Puff loves scritches, all down the side of her short, thick neck, which cause her to pull back her lips in a cute canine rictus. It is a word she already knows; I've been conditioning it while scratching her for weeks.

I summon her over. "Press," I say and point to the button on the kitchen floor.

When Puff was a puppy, I taught her to ding a bell with the cue word "Press." She seems to remember it. She reaches out one paw and presses.

"Scritches," comes my voice from within.

"Yes!" I cry, "Scritches! Scritches!" and give her a prolonged, triumphant scratch. Puff looks pleasantly surprised.

I stop.

She looks at me.

"Press?" I try again.

She reaches out and presses the button.

"Scritches."

"Yes!" I repeat the procedure.

"Scritches." "Scritches." Of her own accord, she presses the button a third and fourth time and receives the appropriate, blissful reward.

Is this a coincidence? I wonder. Is this stubborn little bully really understanding within just a minute or two that pressing *this* button results in *that* desirable outcome?

My question is answered when Dolly, who's been sleeping in the living room, wanders in past Puff's new button. She wags her tail innocently. *What are you guys doing?* she's saying.

With a snarl, Puff leaps at her. *Stay away from my language* appears to be her reply.

Dolly deposits herself, with an injured plonk, in a kitchen corner.

Later, when Puff goes out for a car ride, I decide to try Dolly, who loves nothing more than to sit, muzzle extended and eyes closed in rapture, as a loved one kisses her nose, her ears, her forehead.

"Touch," I suggest, placing a second button on the floor.

She touches it first with her paw, then with her nose.

"Kiss," my voice declares from inside it.

I deliver her reward.

"Touch," I repeat.

"Kiss," says the button.

I offer Dolly a kiss, but she's too busy staring at the button, a bit like I would were a genie to emerge from a Coke bottle.

"Dolly, touch!" I encourage.

With one great Scooby-Doo motion, she leaps away, positions herself behind me, and peers out at it from between my legs.

"Kiss." At arm's length, I press it myself and provide the corresponding kisses. "Kiss." "Kiss." "Kiss."

She accepts them, but without her usual aspect of bliss.

I offer it again. "Want to touch?"

Dolly does not want to touch. She heaves a sigh. Her tail wags, but clearly to the anxious left. I slide the button toward her. She averts her gaze. I leave them there in the kitchen, canine and button, neither talking to the other.

This proves the last time Dolly consents to ever go near a button again.

I know we could do it. I could convince her to change her mind about the buttons. Change that anxious left tail wag into a happier right. With practice and perseverance, we could end up with dozens of usable buttons. When Puff gazed at herself in the mirror, had she buttons, she might also have pressed "WHO THIS." I could even, I'm quite certain, teach the piglets.

But conversation is about finding what's of value to both partners, and, as for Dolly, the buttons aren't for me. I don't want to hear still more of my own voice. I'd rather know what my animals are saying to me in their own natural ways. Maybe meeting the chimps has turned me off forever the idea of teaching an animal to speak Human.

Still, I find the buttons fascinating. I attend a FluentPet boot camp, where I learn from encouraging anthrozoologist Ashley Evenson, that to teach abstract concepts, you must model them, combining words with action. "To teach 'Love You,'" she advises, "say the phrase when your pet is happy or excited; associate it with positive experiences between the two of you. Whereas to teach 'soon' or 'later,' I strongly encourage you to decide what 'soon' or 'later' means to you and stick to it."

For Gal, I consider, "soon" might mean in the next week or so. For me, it most often means within the next five, but preferably two, minutes.

I can see why this might lead to canine confusion.

Button boards are controversial. Proving an animal and their human understand the same thing when it comes to a "ball," "bed," or "basket" is easy; proving likewise for concepts like "love" or "tomorrow" is challenging, perhaps impossible. You can't, say critics, show their workings with science.

I call Leo Trottier, FluentPet's founder, to chat about this. "Traditional science," he says, "values replicable, repeatable results. But FluentPet is 'citizen science,' performed at home, at leisure, under ever-changing circumstances. And biology remembers." He quotes Heraclitus: "'No man ever steps in the same river twice. For it's not the same river, and he's not the same man.' A dog won't respond the same way twice to the button board because she won't be quite the same Dolly or Puff she was before. It might be hard or even impossible to prove they understand the meaning of more complex, abstract buttons when their understanding itself is ever-changing—but that doesn't mean they don't."

Though I consider these AIC buttons a kind of bridge, like Andvari's "go" button or Loulis using the ASL sign for "chase," others deem them a barrier.

"The only provable successes with the buttons," I remember canine scientist Alexandra Horowitz telling me, "have been to show that dogs can press them to tell us something we already *know* they can tell us: 'There's a noise'; 'I want water'; 'I want to go outside.' My dog already tells me these things with where he looks, with his body language."

Having them push a button to do so, she said, is not conceptually richer.

"Actually, it's less good in my mind because you're not looking at the dog's behavior, and you're preemptively saying what the dog means. Pressing 'Outside' could mean 'It's been a long time since I went outside,' or 'I want to go outside,' or 'I *need* to go outside.' But you won't see any subtle differences, which you might see behaviorally, because you've collapsed them all into one button. It's a reduction in understanding the dog's communicative range."

A recent study of button boards in the journal *Animals* makes for fascinating reading. Online videos of "talking" dogs, the authors say, are selected and edited; each narrative is sculpted by a human and, like any communication, tells you primarily what this sender wants you to know. Often, they add, these narratives contain discrepancies between the button-based message and the body language the dog is exhibiting. And there's a crucial paradox, the report suggests, in trying to prove canine consciousness by way of button boards: "The belief," it reads, "that only the acquisition of a human skill, that is, of language, can serve as proof of the mental powers that predate the skill."

Dogs, Alexandra concluded, are anxious to talk to us. "There's so much, behaviorally, they can tell you about their internal states. If we limit their world to this intermediary of buttons, they'll figure something out. But there are so many other ways to talk to one another."

Dolly, for one, would agree, though I doubt she'd press a button to confirm it.

It's a bitter January night, and we are at a magic show at our little local theater. The kids are home from university for the weekend, one with a boyfriend, so there are eight of us altogether, watching the magician onstage eat razor blades, read minds, and make cell phones vanish then reappear inside sealed envelopes.

On the way home, we discuss the show.

Six of us, it transpires, are amazed and curious. How did he do this trick or that trick? Isn't it cool, someone says, that he can do those things at all? Where did he learn? Could *we* be as good if we practiced?

The seventh wants to believe in the magic of it all and will discuss it no further.

And the eighth dismisses the act altogether. Rubbish, they say. Silly tricks. Easy to see through. They spoil it for everyone.

In science, the animal-language pendulum continues to swing. Some still listen in laboratories, others in the wild. Some listen for voices yet to be heard or to those we still haven't wholly understood. (I'm reminded of legendary biologist E. O. Wilson, who found fire ants combined pheromones with other odors to create what he called "proto-sentences.") Others have a vested interest in animals not communicating with humans at all.

"It's convenient to think animals don't have much going for them if you're going to put them in a lab and experiment on them," Con Slobodchikoff told me. "And then there are people who've made a career out of saying that animals can't do this or that, and it's hard for them to turn 180 degrees and say, 'Oops—actually they can.'"

Some, like Mary Lee, think we should never teach animals to speak Human again. Others keep trying. In 2018, at the Marineland Aquarium in Antibes, researchers successfully trained a captive orca named Wikie to imitate human speech through her blowhole. A sort of sad, cetacean Viki, she can say "Hello," "Bye bye," and the name of her trainer, "Amy." It's not clear, though, if she knows what they mean.

Snow drifts silently over the farm. Icicles form on the downspouts, carved into weird diagonals by the north wind. Under a sky the color of dirty bedsheets, the horses stand shoulder to shoulder, tucked tight into the lee of a pine, motionless, expressionless, silent, battened down.

One dark, nasty day during January's frigid finale, I sit huddled by the fire as the mercury retreats, like a groundhog, to fifteen below.

I cuddle with Puff and Dolly and think of Nim and Viki and Wikie and read more about that nineteenth-century Nature Fakers battle, which in many respects is still raging today. Both sides zinged essays back and forth, critiquing each other. Scientists versus naturalists. Impartiality versus individuality. Scientific method versus in-situ observation. "Say what you see" versus "interpret what you think you see." Tensions escalated. Teddy Roosevelt, an avid hunter, gave Ernest Seton's camp the "nature-fakers" moniker. Jack London responded, calling Roosevelt "homocentric." I find a 1920 verse titled "Proof" by poet James J. Montague:

> John Burroughs, who's a shark on birds
> (He classifies 'em by a feather),
> Avers that they're devoid of words
> And simply cannot talk together.
>
> He gives the nature-fakers fits
> Who picture birds in conversation,
> And tears their story books to bits
> In scientific indignation.
>
> But there's a wren outside my door
> That talks whenever I go near him,
> And talks so glibly, furthermore,
> That I just wish that John could hear him.
>
> Of mornings, when I stroll about,
> The while he hymns his glad thanksgiving,
> He interrupts himself to shout:
> "Hey! Ain't it glorious to be living?"

What, I consider, have I learned from all this scientific diversity? To listen with my nose and my eyes and my instincts as well as my ears. Not to underestimate the small, the strange, the silent. To hear both soloist and symphony. To watch and wonder.

Science lays groundwork. It grapples with big questions. It looks at both the patterns and the anomalies. It knows what osmosis really is.

I've learned that a party full of animal scientists would be a fun place to be—unless they went for one another's throats. That in a glimpse of commonality, you can find communion. That the individual and their umwelt might be an insoluble puzzle, but that we should still try to put ourselves in their paws or claws or tentacles. That complex speech isn't a precondition of having something meaningful to say. That there's never one right answer, but that always, *it depends*.

In 1665, Dutch physicist Christiaan Huygens discovered "entrainment" with pendulum clocks. He set them in motion at different times and found the next day that the pendulums had synchronized their swinging. Despite all our differences, whether we're an ethologist, biologist, chimp, cat, fruit fly, or even a linguist, one thing makes our pendulums swing in synchrony: We all talk to be understood, but we listen to try to understand. The differences make the conversations interesting. The similarities make them possible.

For most of my life, I was scared of science because of what I thought was its certainty. I loved literature, poetry especially, for the opposite reason. But though there will always be those who guard language every bit as viciously as Puff ("It's ours," they say to animals, "and you can't have any"), I've found science is full of wonderful, thoughtful questions: stuffed with what-ifs and can-theys that have led to some truly remarkable interspecies conversations. My favorite animal scientists are not certain at all. "If you're sure something doesn't exist," Irene Pepperberg advised me, "wait another decade or so, and you'll probably be proved wrong."

I get up from the fire to fetch the book of the poet Keats's letters I gave myself for Christmas. I flip to a passage I've underlined with thick pen. "Negative Capability," Keats wrote in the winter of 1817, "that is when man is capable of being in uncertainties, Mysteries, doubts, without any irritable reaching after fact & reason."

Good science and good poetry have something in common after all, I decide. Employing some scientific negative capability opens our ears and eyes and noses to hear, to know we don't know what we don't know, and be OK with that until the day we find out.

I snap Keats shut and get up to make the horses' supper. The science of communicating with animals, I conclude, is not a closed book; it's a story being written.

And so are they.

And so are we.

By 2:00 a.m., the temperature has dropped to a cool minus twenty-five degrees, minus forty with the windchill. Gal vanishes from bed to check on the piglets. He's gone for an hour.

"Where were you?" I murmur as at last he slips back beneath the covers, toasty warm and with a summertime scent of hay.

"They were pleased to see me," he says, "They invited me in. So I stayed awhile."

PART II

The Doers

The Practice of Listening to Animals

train · ing /ˈtrāniNG/

noun

the action of teaching a person or animal a particular skill or type of behavior.

6

Maybe She's Thirsty

Learning to Listen, Not Tell

> "First you go with the horse. Then the horse goes with you. Then you go together."
>
> —ATTRIBUTED TO TOM DORRANCE,
> legendary horseman, 1910–2003

FEBRUARY

Dark days in deepest winter. Snow burdens the branches of the oak; they creak, rheumatic, outside our bedroom window. A sudden thaw followed by a terrific ice storm cause the air to crackle with the static of a billion flash-freezing raindrops, a sound so eerie it makes my skin prick with goose bumps. Soundscape ecologist Bernie Krause, I think, would enjoy this supernatural geophony, but I'm too spellbound to pull out my phone and record it.

Each day, I struggle out before dawn, delivering forty-pound hay bags like some flimsy Father Christmas. I Jack-Torrance the horses' frozen drinking water. I weep at the glacier still busy conquering their corral. They'll kill themselves on it, I fret. A broken leg is a death sentence for these half-ton concrete blocks supported by the slenderest porcelain columns.

I bring a shovel blade down hard on the ice. The sound reverberates, metallic, across otherwise silent fields. I try again and again till I can feel it in my teeth. "Fuck you, weather!" I yell as I barely make a dent.

My friend Natalia, an unsentimental horse-owning engineer, tells me not to worry.

"Horses," she scolds, "are built for this. If not, then it's Darwin Awards, Equine Edition."

This makes me feel a bit better. I buy ice cleats for my own two feet and resolve they'll be OK on their collective eight.

Reluctantly, I also postpone this year's horse-training goals till better weather. I always maintain a rolling list of things I plan to teach, a mix of useful skills and fun party tricks I hope we'll mutually enjoy, but it's too cold for this weakling human, even bundled up, to do more than keep the horses alive. Instead, I decide, I'll delve deeper into the nuts and bolts of animal training itself. What happens when we replace its traditional *telling* with *listening*? Which techniques improve that listening, and what can they help us hear? How does animal training differ from animal science? What more is there to discover in this sphere already far more familiar to me?

"The science people," Jenifer Zeligs, an animal scientist and trainer who we'll meet a little later, tells me, "are listening to animals for the sake of learning: to know more about ethology, natural history, linguistics. The training people are interested in its practical applications. They want to *do* something with that listening."

Not everyone's a scientist. But anyone with an animal can be a trainer. You don't need to go to university for it. If you've inclination and a guinea pig, literal or figurative, you can train them. Naturally, there are also a thousand courses you can take and certificates you could attain. But to start, all you need is good listening, good timing, plenty of patience and practice, and an understanding of the underpinnings of some basic techniques, which, I'm to find, are all firmly couched in science.

Yet unlike scientists, trainers are free. Call something "speaking" or "talking" or talk about what an animal told you, and no one but the internet trolls will dispute you—at least publicly. You can experiment without writing it down. You don't have to publish your findings. You train what you feel like training, where you feel like training it, when you're so inclined. For most people, training is not a paying profession, but it doesn't cost much either, except for your time.

So why do it? Training helps our pets become solid citizens with life skills to succeed in our human world. It is vital for service animals and for those with jobs to do. Training can lower the stress of routine medical procedures for captive wild animals and provide them with enrichment. A 2017 Durham University study on zoo-kept ring-tailed lemurs found training increased their affiliative behaviors both toward human keepers and also, outside sessions, toward one another. And undertaken through listening, training is fun for everyone, building trust between trainer and trainee.

I mull all this over on these ferocious February mornings, as the oak-tree starlings pipe and whistle me on overhead. As eminently trainable as other wild birds more commonly kept as pets—mynahs, for example, or the mobs of parrots and cockatoos I've watched flying free in Australia—it has always struck me as strange that people don't love starlings more. Because the humble *Sturnus vulgaris* is largely unbeloved by people; the name alone sounds vaguely insulting. They're too ubiquitous, we humans say, too destructive, too messy, too noisy. Accused of killing others of their species, of crowding them out, starlings are the ultimate survivors, inhabiting every continent except Antarctica. It's almost, I consider, as if we were talking about us humans. ("Everything that irritates us about others," wrote Jung in *Memories, Dreams, Reflections*, "can lead us to an understanding of ourselves.")

But to me, their songs are magical, as is their apocryphal backstory. The first batch, it relates, was brought to North America by Eugene Schieffelin, an eccentric ornithologist who tried releasing every bird referenced by Shakespeare into New York's Central Park (starlings, however, receive just one single mention: in *Henry IV, Part 1.* "Nay," says Henry Hotspur of the king. "I'll have a starling shall be taught to speak nothing but 'Mortimer,' and give it him, to keep his anger still in motion.")

They are also excellent imitators, a skill that found favor with another monarch: King Charles II, who trained his pet starling to speak. "One day ye Bishop of Canterbury came into ye bed chamber," recounted Sir Francis Fane, court playwright, in 1672, "& ye bird hop on his shoulder & sade 'Wilt thou have a whore, thou lecherous dog?'"

I'm with Charles. My spirits rise when I open the door to the starlings' wild jubilance, their *whirrs* and *pheeeeeeews* and *wee-wee-wees*, while the wind howls and the sky is limned peachy behind them. And though, unlike the unfortunate bishop, I don't know what they're saying, I take to talking to them anyway, offering up a daily "Good morning, Mortimers," to gently train them to anticipate my presence.

On less blustery days, Indy and I strike out along the rail trail, tracing stories written in footprints. Our eyelashes frost together when we blink. Puff bounces beside us through the drifts. We follow deer tracks and vole tracks and pheasant tracks, or rather, *a* deer's tracks, *a* pheasant's tracks, *a* vole's tracks. Here went one individual vole, burrowing her perfectly tubular tunnel in the snow. There went one certain red squirrel—Sven, perhaps—from tree to tree, his tracks skipping a couple where he traveled up and over. The pheasant prongs belong to Siegfried or one of his ladies, the smaller prints to our pigeon pair. And those pad marks the size of a child's hand? The bobcat we captured on camera, out to make a meal of one of these other individuals. Our footprints mingle, our lives complect.

Occasionally the tracks are fresh, and we spot their owner—a deer, usually—up ahead. Conscious of our body language, imagining their umwelts, we practice getting closer. We find the direction of our feet matters; predators' feet must face their prey. So we train ourselves to advance casually, sideways, what Indy calls our "Doopty-doopty-do." We watch sideways too, like villains in a Victorian melodrama. We start to know if we still have wiggle room. We learn to feel for that space, to try to expand it, like gently blowing a soap bubble. Push too fast, too far, and pop, it's gone.

Still, it's not easy. Human nature always wants a little bit more. To us, proximity equals connection. But Sedna the octopus reminded me nothing good comes of grasping. We're not butterfly collectors. Our eyes and ears are not a greedy net. We learn to be happy with what they offer, not what we want.

"Often," primatologist Cat Hobaiter told me, "researchers habituating a group of chimps will follow them, and when they lose them, that's the end of the day. But I always try to do the thing that's very difficult: I find the chimps, spend a bit of time with them, and if I think it's going well, I walk away. That's such an important part of building their trust—that I always leave on a good note.

It's not always easy, she admitted. "Because often you've worked your arse off to find them. In northern Guinea, we might walk twenty miles a day across high, open savanna, in tough terrain and 104-degree heat. We might've gone for weeks without seeing them and have finally found them, and we have to be cool and walk away."

I was startled to hear this, because it's so like my own kind of animal training: "Quit while you're ahead" and "End on a high note" have been my watchwords in training sessions, hard as they can be to stick by, for years. Though I'd always thought I was afraid of it, it turns out I've been assiduously employing science all this time.

On February 14, we spot Siegfried's face, Valentine vivid, peeping out from a snowbank.

We sidle, stop, sidle, until we're fifteen feet away and something inside him seems to tighten.

So we stop and stay.

And so does he.

We stand with Siegfried until snow gets into our eyes and our fingers grow numb and he goes back to pecking, then head home to defrost by the fire.

It's a shame, says Indy, we can't invite him in with us: the Pheasant Who Came to Tea.

On February 18, a storm knocks out the county's power. I light candles and break out a bag of cheesy puffs to teach the dogs some silly new tricks. In four five-minute sessions as the storm furies and candles flicker, I teach Dolly to roll back and forth like a furry windshield wiper and teach Puff to balance on her haunches in that classic begging pose.

Animal training is always an enriching conversation. I ask a question: "Would you like to try this?" I listen to the reply, sense when the trainee needs a break, notice their moments of joy and keenness and distraction and puzzlement, of "Is *this* what you mean?" and "Ohh, *now* I get it," and "I'm sick of this," and respond accordingly.

All animals like learning, finding out, and, ideally, succeeding; at its most basic, that's what drives the survival instinct that keeps us all alive. The great thing about training is that the trainer is learning too, and learning together makes every small success doubly rewarding. They say if you want to know how much you don't know about something, try teaching it. Training, I've found, offers excellent opportunities for measuring your ignorance, then trying to do something about it.

For centuries, though, this exciting, collaborative process has been overridden by the "Sits" and "Stays" and certainties of telling. "Show him who's boss!" was the maxim of every rod-backed 1980s riding schoolmistress I encountered as a child. "Don't let him get away with it!" Smack the dog when it chews your sneaker. Rub the cat's nose in it when it pees on your bed. If the horse won't go, insist. If it still won't go, *insist louder.*

But there have always been exceptions. "What a horse does under compulsion," wrote philosopher and horseman Xenophon in the fourth century BCE, "…is done without understanding; and there is no beauty in it either, any more than if one should whip and spur a dancer." Cheer on your hunting dogs, he instructed in his treatise *Cynegeticus.* "Keep up their spirits with a constant 'Well done, good hounds!'"

"Never forget," echoed Major-General W. N. Hutchinson in 1847, though his book was called *Dog Breaking*, "to have some delicacy in your pocket to give the youngster whenever he may deserve it."

I teach the dogs our storm-day tricks using these same two positive-reinforcement techniques of praise and food rewards. I cluck my tongue at the precise moment Puff exhibits a step toward the behavior we're learning, then pop a cheesy puff into a happy doggy mouth. This cluck is a "bridge signal." It means "Yes! That's right!" and shows your learner, so long as your timing's impeccable, that *that* action, exactly *there*, is what you were asking for, and that reward is on its way. Also known as clicker training or R+, it's been used for decades to train dogs, marine mammals, and performing animals and forms one of the four quadrants of "operant conditioning"—a term invented by that infamous, pigeon-boxing behaviorist B. F. Skinner.

Operant conditioning is, in a nutshell, *active learning.* Your learner does something and receives a consequence. Generally, if that consequence is good, the behavior—whatever it was they just did—will increase and vice versa. (Though it's important to remember it is the learner who decides whether the consequence is good or bad. For a human toddler, for example, any attention, even being scolded, might be more rewarding than no attention at all. Conversely, for some timid animals, a congratulatory pat might feel like a punishment.)

Within this basic framework sit those four quadrants. Think of a square, divided into quarters. In two boxes on one side sit positive reinforcement and negative reinforcement; in the other two, positive punishment and negative punishment. All training—of aardvark, cockroach, human being, or zebra—encompasses at least one, often more, of these quadrants.

They sound daunting, but aren't really. Positive reinforcement simply means the learner receives something pleasant—an *appetitive*, like a cheesy puff, is *added*—as a result of their actions. You get a math question right, you're awarded a point. You do a good deed and receive a thank-you. Your puppy sits when requested and gets a cookie for his troubles. Negative reinforcement (or R-), meanwhile, simply means the *removal* of something *aversive* as the consequence of an action. You tug on your dog's leash to ask them to come to you. They start coming. You stop tugging. You squeeze your feet against a horse's sides to request he walk forward. He starts walking. You stop squeezing. Positive and negative punishment are equally straightforward, though with animals generally less effective. A positive punishment might be a policeman *adding* a speeding ticket if you're caught driving too fast; a negative punishment might be *removing* a teenager's phone for not putting it away at the dinner table.

It occurs to me again at this point what an awful lot of science is folded into training. Because really, I realize, science is part of everyday life: every recipe you bake, every time you try something out, test it, refine it, and try again, you're following basic scientific processes. I feel silly for being scared of it all this time.

The second core concept in animal training is classical conditioning, or *inactive learning*. It's the prototypical Pavlov's dogs: A neutral stimulus (in Pavlov's case, a bell ringing) is paired with a positive one (the dogs' dinner arriving) until that originally neutral stimulus takes on a new, positive meaning. Unlike operant conditioning, the dogs themselves didn't have to *do* anything. They just heard the bell, associated it with the imminent arrival of food, and began to dribble. Classical conditioning helps me perform a little Pavlov myself by transforming a neutral stimulus, that "cluck" sound I'm making with my tongue, into the bridge signal that means "Excellent job! Cheesy puff incoming!"

But what else do you need, aside from some operant and classical conditioning (which we employ all the time, with both humans and animals, without even knowing it), to be a good trainer? In some ways, I reflect, as the dogs and I trick-train the diabolical afternoon away, good training is like being a solid human friend. You try to be a careful listener. You try not to presume or pass judgment, not to put on a facade. You find opportunities to have fun together. And when a difference of opinion or understanding surfaces, you attempt to handle it gently, skillfully, not come on too strong or impose your will for the sake of being right.

Dolly has had enough of her new trick. I give her an extra-large "jackpot" treat to proclaim the session over, put away the cheesy puffs, zip up my coat, and head out into the storm to feed Constance and Agnes their dinner.

The piglets, now eight months old, have turned into swineagers. Appalled by the weather, they stay in their bedroom, emerging only to eat. They are pleased when I come bearing gifts: turnips, rutabagas, and the black-oil sunflower seeds that act like an internal ChapStick, keeping their skin from turning flaky and dry. They're also happy with a visit for visit's sake. I lie on their edible mattress, scratching their bellies, their cheeks, beneath their chins; Agnes likes it if you gently hold her trotter as if it were a small child's hand. I visualize the burst of oxytocin racing around my bloodstream and theirs. Some studies I've read suggest cows enjoy listening to classical music. I'm not sure about pigs, but I sing to them anyway and chat in low, comforting tones. According to my pigtionary, their response is one of contentment-slash-delight.

Now and then, when the sun comes out, so do they, dragging their hay outside to lounge like old ladies on Brighton Beach. Though each has grown by about forty pounds and been trained to observe a few human-beneficial manners (Agnes now waits while Constance receives her food bowl first; Constance has learned she won't get dinner till she stops jumping up), their personalities haven't changed. Constance is still gregarious and cuddly and bossy with her sister. Agnes is still more reserved, interested in affection only on her own terms. But she often watches when I leave, emitting a plaintive little *wheee*, whereas voluptuary Constance just goes right on back to breakfast in bed.

I consider how these differences might influence my plan to train them this summer to walk on a leash. I imagine Constance will be easy—that I can just pop on her harness, jiggle a feed bucket, and away we'll go. Agnes might take more warming up to the idea. I might have to go slower. I might have to consider her mood: the individual, in the moment, within the species.

But is there anything I'm forgetting?

I decide to ask Dr. Jenifer Zeligs, whose textbook, *Animal Training 101*, details the pros and cons of what she calls "the many strategies of passing meaning between one another."

Jenifer is an unusual combination. With a PhD in biology and a background in psychobiology, physiology, and behavioral science, she has been a real-world animal trainer since 1979 and is also on the path to ordination as a Buddhist priest. Though she's trained everything from porcupines to rhinos, she has a particular penchant for sea lions. "They are," she tells me, "both marine and terrestrial so have an 'everyman' view. They're cognitively advanced, in the same canid group as wolves and dogs, with the potential to develop an extremely elegant, almost Dr. Dolittle level of interplay with a human that looks from the outside like mind reading."

A quick YouTube search turns up an old clip of Jenifer on *The Tonight Show* with her rescued sea lion, Sake, who balances a ball on her nose, performs a handstand, and plays catch-and-toss with a Frisbee for a constant stream of seafood rewards.

Above all, Jenifer tells me, it's crucial to remember that animal training encompasses "*all* behavior modified in the presence of a human being."

In other words, training isn't what you *want* to happen, it's what you're *causing* to happen. "Nothing means nothing. Everything means something. You're training even when you don't think you are and don't mean to be."

This reminds me of Clever Hans, the famed early-twentieth-century German horse taught by retired schoolteacher Wilhelm von Osten to solve complex mathematical problems. Hans invariably provided the correct answer by stamping one hoof the appropriate number of times, "according," said the *New York Times* in 1904, "to the same psychic laws as we use a language to make others understand." Then one day in 1907, psychologist Oskar Pfungst caught Hans out: The horse, he found, was simply reading humans' unconscious body language. As Hans approached the correct answer, his questioner's face would tighten unintentionally with tension. When he reached it, that face would relax, cueing Hans to stop tapping.

But Hans, I believe, wasn't a fraud—just a better listener to humans than most humans are themselves.

Second, Jenifer tells me, "the very best animal training is a conversation between two willing participants."

To this end, she says, begin by building a relationship with the animal you're hoping to train, be it penguin or puppy. "Be generous. Make a connection before you give a direction. Come bearing gifts that are noncontingent, based only on your wish to demonstrate goodwill." With a sea lion, bring fish. With a camel, carrots. Both are deposits in what she calls "the relationship bank account."

"Next," she tells me, "allow the animal to direct the conversation before getting into whatever your own thoughts or desires might be." Hang out for a bit; see what they'd like to say or do rather than drill-sergeanting your way into the encounter.

And throughout, Jenifer advises, listen deeply: to their sounds, their facial cues, their body language and gestures, and always remember that, like Hans, they're listening to yours too.

"From an earwig to the Dalai Lama," she tells me, "everyone on the planet is simply trying to be happy. So to train someone to do something you'd like them to do, you need to think about how to make it something they will like too."

The conundrum, though, is that emotion and behavior are ever-changing. "Professor Susan Friedman says, 'Behavior is the study of one.' But beyond that, it's the study of one in one second, in the right now." Like Alexandra Horowitz, I recall, with her dogs, or Heraclitus's river. "You've got to listen to the incredibly rich soup of motivators that change in your learner from moment to moment. And that takes an open, curious, spacious, nonjudgmental "beginner's mind." In Buddhism, we'd say, 'In the expert's mind, the possibilities are limited, and in the beginner's mind, they are many.'

"Therefore it's important," concludes Jenifer, "to surrender to the fact that whatever you know is approximate. It's a best guess, a high-percentage chance. And that's the magic, right there: to stay in your deep listening, with a curious beginner's mind and an understanding of these basic training tools, and remain fluid to the evolving situation. I heard a great quote recently," she adds. "'An expectation is a resentment waiting to be born.'"

Food for thought, before next summer's Project Piglet.

It was about a decade ago, high in the Chiapas mountains, that three words changed the way I approached horses forever.

"Maybe she's thirsty," said Sam, a Jeff Bridges look-alike, of the sweet little mare with whom I was trying and failing to communicate as we prepared to depart for the hills.

I felt something inside me slip sideways. *Maybe she's thirsty.* It was so simple. Why hadn't I thought of that?

The answer, I realized, lay in decades of classical and operant conditioning.

In 1980s England, where I first learned to ride, horses were what they were. Riding-school ponies were stubborn and disobedient, because that's what ponies were like. Thoroughbreds were born for racing, because they couldn't stand still if they tried. There were draft horses who plodded along pulling drays because they were big and strong and not very bright. And police horses, tall and fearless, who didn't mind plowing through poll-tax rioters or drunken football hooligans. As pupils, we did what we were told, which was to make the ponies do what they were told, both with the looming threat: *or else.*

Later, I rode wherever I could. At Petra, in Jordan, I galloped a Bedouin racehorse past astonished tourists. I rode through Israel's Negev desert for *National Geographic* and through Belizean jungle for CNN. I loved riding. I knew how to tell a horse what to do and what not to. I was gentle. I never smacked or kicked or shouted; I always declined a whip or spurs. But I wasn't worried about listening or our relationship. "Good boy," I said, hopped on, and off we went.

Now here I was in the Mexican highlands, unaware everything was about to change.

I took hold of the little mare's lead rope to bring her out of her pen.

"Walk on," I said, and merrily walked forward.

The mare did not.

I turned. "Walk on," I urged.

Nothing.

I moved a few steps back, held the lead rope tighter beneath her chin, and tried again.

The mare, in a quiet, equable, and utterly good-natured way, did not take a step. She didn't *react*. She just didn't *act* either.

And then the three words happened.

"Maybe"—Sam gestured to a water trough—"she's thirsty."

I blinked as if he'd just reached out and flicked me between the eyes. Maybe she's thirsty. It wasn't that I didn't think horses *could* be thirsty. But despite seizing every opportunity to spend time with them, it had not occurred to me that a horse might want to tell me about it.

I felt mortified.

"Um," I mumbled. "Do you… Do you want a drink?"

I loosened my grip on the lead rope, and mildly, with no fuss or pulling, the mare led me to the water tank, where she dipped her head and took a long, deep draft. She wasn't being difficult or disobedient. She wasn't even saying she didn't want to do what I asked. She was simply requesting, quietly and politely, "Could I please have a drink first?"

That night at our rented apartment, beside a tiny front-room chapel where a congregation sat worshipping under fairy lights, I asked Google, "How do you listen to your horse?"

In reply, Google sent me an Australian horseman named Warwick Schiller.

The following year, Major came into my life: a sizzling red fire-cracker who had bucked off a professional bronco rider and been returned by several prospective buyers.

But thanks to Warwick, I was ready and listening.

"You can be a functional animal trainer without being a good listener," he tells me today, years later, by Zoom. "But you'll never be a great one."

Warwick is just back from riding the Gaucho Derby, the "world's toughest horse race," a ten-day, three-hundred-mile jaunt across the wilds of Patagonia.

"And for years," he continues, "that's what I was: functional."

Functionality served him well. Warwick grew an online following. He taught clinics all over the globe. He helped thousands of people with their horses' behavioral problems.

Then, he says, he met a horse named Sherlock to whom the rules didn't quite apply: an aptly named individual (I remember Isaac and his *Sign of Four* cockroaches) for sending him on a journey of discovery.

"Sherlock was highly trained yet with some small issues I just couldn't fix. I've never had much self-confidence, so when something doesn't work, I tend to get curious."

This reminds me of the perils of presumption. From the outside, Warwick seems self-assured, poised, unflappable.

"I stopped trying to change him, allowed Sherlock to be who he was, and looked around to find what else was out there: what I didn't know. I didn't start out *trying* to listen, but my innate curiosity got me to the point where I realized I wasn't.

"Then one Saturday night, in the midst of a three-day clinic, I had a nervous breakdown. Or, as my therapist would call it," he smiles, "a 'spiritual awakening.' I woke up next morning and thought, 'I don't know if I can go out in front of these people, because I don't know what the hell I'm doing.' I had to tell everyone: 'Bear with me, because I'm going to try some new stuff with these horses.'"

The stuff he tried, he says, "all came down to listening."

Such as to Cody, a nine-year-old mustang horse with a problem.

"Occasionally, out of the blue, Cody would bolt away with his rider. She wasn't sure what caused it or how to stop it happening. Instead of trying to solve it, I noticed his anxiety, the small signs of concern he showed every time I tried to approach him. I let him know I saw them, and in reply, I backed off."

Cody's signs of concern, like all horses', were subtle: a slight turning away of the eye, a tensing of the nostrils or the jaw. A change of breathing, a minuscule raise of the head: signals Warwick calls "stress indicators" that are always, he says, "important information."

"By showing him I noticed them, I let that horse know he was heard and therefore safe."

The results were dramatic. Cody lay down and slept for hours in the middle of the arena, in front of dozens of spectators and other participants. He had never slept in a human's presence before.

"I didn't know why or how it worked at the time, but the more I thought about it, the more I realized it was because I was communicating my awareness of *his* communication of his concern, which made him feel safe enough to lie down. I was also communicating my willingness to allow him to have a say and to hear whatever his answer was to the question 'Can I come closer?' There was no judgment. Just curiosity."

Cody, Warwick tells me, never bolted again.

"Why," I ask, "does listening instead of the customary telling work so well as a basis for training?"

"Traditional training," Warwick replies, "is based on the expectation that since you spend time, money, and energy on maintaining your horse, you are entitled to certain things in return.

"But if you come from a *giving* place instead of a *trading* place, the whole thing's completely different. Listening allows you to come from that giving place."

Warwick's journey led him to neurobiologist Stephen Porges's polyvagal theory, which proposes that "when humans feel safe, their nervous systems support the homeostatic functions of health, growth, and restoration, while they simultaneously become accessible to others without feeling or expressing threat and vulnerability."

Connectedness, explains Porges—being accessible to others—is a "core biological imperative."

The human autonomic nervous system, Porges says, consists of two branches: the "activated" sympathetic system, responsible for our fight-or-flight responses in the face of threat or danger; and the "lower-activation" parasympathetic system, which has two further branches: a calm ventral vagal state, where humans experience feelings of safety and social connection, and an immobilized dorsal vagal state, where we react to stressors by freezing, shutting down, dissociating. Humans share their basic autonomic nervous system with all mammals, meaning many mammals too find their sense of safety—and, through it, connection—via the parasympathetic ventral pathway.

"Polyvagal theory," Warwick tells me, "explains why my listening allowed Cody to feel safe. It's the science of connection. We feel connected through feeling safe. We feel safe by feeling listened to. Someone communicating that they're listening to us gives us all a tremendous feeling of safety."

After Cody and Sherlock, Warwick experimented further and started to notice consistencies.

"I wasn't trying to prove listening was 'the thing.' Listening just kept showing up. I realized that everything that worked fell under this umbrella of communicating to them that I was listening, that my attunement was what made them feel safe. And when horses feel safe, a lot of what we call behavioral problems just aren't there."

Warwick uses clinical psychologist Richard Erskine's definition of attunement. It is, Erskine said, "a kinesthetic and emotional sensing of others knowing their rhythm, affect and experience by metaphorically being in their skin, and going beyond empathy to create a two-person experience of unbroken feeling connectedness by providing a reciprocal affect or resonating response."

Though Erskine was describing human therapist-patient relationships, the definition applies just as readily to trainer and equine trainee.

Most horse-training techniques, Warwick says, are for horses who have never experienced such connection with a human and as a result have dysregulated nervous systems.

"If you get rid of that dysregulation, and they're as relaxed hanging out with us as they are with their friends, it's a totally different ball game. Then if a problem does appear, you can ask yourself, 'What happened before the thing that happened happened? What tiny signs of concern did I miss?' It's a mindset, not a method."

The effects of this mindset, Warwick tells me, extend to humans too.

"People tell me frequently that this approach has changed not just their relationship to their horse, but every part of their lives. It's very profound. Often they'll visit my booth at a horse expo and burst into tears. I listen and give them a hug."

And listening to the tiniest communications, Warwick says, works with other animals too. Out hiking in Australia, he came across a kangaroo. "Normally when kangaroos see you, they freeze, and then they leave. So I froze too. *I see your concern*, I was saying. He stayed and went back to grazing."

It makes me think of Wordsworth, I tell him, at Tintern Abbey:

While with an eye made quiet by the power
Of harmony, and the deep power of joy,
We see into the life of things.

One snowy February night, another wicked nor'easter comes barreling in, making the wood roar with indignation, bowing the birches, forming a snowbank so deep against the west side of the house that the little chair on which, on sunny days, I sit talking with the piglets is thoroughly buried.

Next morning, the horses enjoy the fresh fallen snow. They roll and buck and kick up their heels, playing follow-the-leader among the stands of holly and wild cranberry.

But by evening, when I arrive with their hot dinner, the wind has picked up, and the fun and games are over.

Major, no longer lame and limping, races into the corral, spins, and races out again.

Andvari skitters, tosses his head, tries to kick him.

Something in the old barn clanks, and both careen off again into the twilight.

This is not bad behavior. This is, as Warwick would say, important information.

Imagine you're in your bedroom, in the dark, and feeling for the light switch. In the corner, something topples: a pile of books from a nightstand or the cat dislodging a potted plant. Do you react as you would if all the lights were on? Likely not. You might, if you're like me, jump several feet in the air, imagine monsters, burglars, something about to grab your ankle from underneath the bed.

Logically, your prefrontal cortex knows this is probably not the case. Still, your ancient, almond-shaped amygdala, the limbic "lizard brain" tasked with keeping you, along with all other mammals, birds, and reptiles, alive at this moment, overrides it.

This is how it is to be a horse in the wind: a prey animal who relies on his sense of smell and acute hearing to keep himself safe. High winds impede these abilities; strange aromas blow in from far away: the "scent of blood on the hills," my friend calls it. Noises are muffled, distorted, accentuated. It's impossible to tell where that ankle grabber's lurking.

I wait until they return, snorting, to the corral. I speak slowly and soothingly as I place down their dinner bowls, hang up their hay nets, demolish the ice crust forming on their water.

"'Who has seen the wind?'" I recite some comforting Christina Rossetti. "'Neither you nor I. But when the trees bow down their heads, the wind is passing by.'"

Then I stand sentinel while they eat. *I hear your concern*, I'm communicating to their sympathetic nervous systems. *But don't worry. I've got you.*

First Major, then Andvari, lets out a deep sigh: a sign each is returning to his parasympathetic ventral state.

"Good boys," I say. "I'll come check on you later."

I leave them calm and quiet, munching softly, as the night bellows on.

On February 25, we flee the snow to visit our friend Peter, a musician and boatbuilder who lives on a crumbling Barbados plantation. His home was once a zoo. I love staying in the former stables and swimming in the old penguin pool while Indy stalks lizards in the iguana enclosure, where a sign still warns that It Is Forbidden to Feed the Reptiles Bread.

I spend the flight south reading up on the methods of Dr. Bob Bailey, an accomplished trainer now in his eighties who trained "spy cats" and seagulls for the CIA in the paranoid, and highly inventive, days of the Cold War. He taught these cats to walk for miles to eavesdrop on enemy conversations via hearing-aid-like cochlear implants, the gulls to retrieve or drop covert devices far out to sea. For both species and any other, Bailey explains, the two keys to training success were the same. First, listen closely to your individual to find out what they find valuable and what will therefore make the task worthwhile to them. And second, work incrementally toward your goal, rewarding each step along the way, "because success itself," Bob says, "is reinforcing," both for the animal and the trainer.

I'm still contemplating this lesson as, under Barbadian sun, we settle in and thaw out. Peter's estate is managed by five zealous border collies and is populated by mongooses, monkeys, house geckos, and scaly-naped pigeons. There are moths the size of dinner plates and beetles the size of brooches. Barbados lies at one end of the geographical length of elastic that sends migrating species hurtling north to south and back again. Mervyn, I realize, might be in this very garden. The grackles bathing in Peter's birdbath might be the ones frequenting ours in summer. Bats flit in and out of the bedrooms. Indy finds a four-inch *Pseudosphinx* caterpillar crawling, bristly and alienish, up the bathroom wall.

Each morning while I brew coffee, the collies arrive for their daily due diligence, sniffing about in the corners. When nobody's watching, I sniff with them too. Historically, Indy has accompanied them on their rounds, but this time, he is busy with a small, spherical house sparrow, the kind birders call a "little brown job," whom he names Falafel.

Indy first notices Falafel hanging about outside the stables. Perhaps he lives nearby, he says, or has a braver personality: "a Rita, not a Mabel." He tosses him crumbs. Falafel swoops down to gobble them. Indy comes up with a training protocol; his goal, to have Falafel literally eating out of his hand.

He moves the next meal to the living-room step. Falafel becomes a return diner.

Slowly, patiently, over the next two days, the pair make their way up the steps and into the open, windowless house. The word "patience," I tell Indy, derives from the Latin *patior*, meaning "I suffer," but in this case, there's no suffering: just mutual, reinforcing delight in their slow, steady progress.

On day three, two further sparrows, this pair more reticent, arrive with Falafel and perch on the back of a corner armchair to watch.

Now here they all sit on this fourth morning, child on sofa, sparrows on lampshade and armchair, as Indy tears bits off his breakfast toast, and Falafel glides down to the coffee table to take them.

"I think he's ready," Indy says.

In the corner, his two audience members flutter anticipatorily.

He places some crumbs in one outstretched palm. Falafel wastes no time. One fat little feather ball flies from lampshade to palm, lands, and starts pecking away at his breakfast.

Dr. Bob Bailey, I believe, would be proud.

It's on the morning of day six that Falafel starts training Indy. He is already waiting, on the kitchen lampshade this time, for Indy to come out of his bedroom. He performs flybys while breakfast is prepared. He starts arriving at other times of day, just sitting, watching, waiting. They move together around the kitchen.

The communication is clear. "You are safe" and "Would you like?" have become "Can I have?" and "Hurry up." Children, I realize, do this so effortlessly, before their listening gets lost in all the noise.

By the time we're repacking our suitcases, Falafel's companions, albeit more flightily, have started joining in. We sing "Three Little Birds" around the house and find we can't stop.

"First you go with the horse. Then the horse goes with you. Then you go together." This elegant adage, from legendary old-timer horseman Tom Dorrance, pops into my mind one lunchtime. The same, I decide as I watch Indy and friends dining together, can be said of sparrows.

Back home at the farm a week later, an email arrives from Peter. THIS YOU? is all it says. Beneath it is a screenshot of a just-posted Airbnb review.

"We had a lovely time at Peter's," it reads, "including feeding the birds at breakfast (they would be waiting for us to get up to be fed)."

Indy is delighted. "Falafel!" he cries. Then immediately, "When are we going back?"

Know Thyself

How Self-Training Helps Us Hear

"Public opinion is a weak tyrant compared with our own private opinion. What a man thinks of himself, that it is which determines, or rather indicates, his fate."

—HENRY DAVID THOREAU, *Walden*

"Two tears in a bucket. Motherfuck it."

—THE LADY CHABLIS, 1957–2016

MARCH

There is a man in the south of England who sends animals to sleep. I've seen him do it: to goats, mules, donkeys, llamas, and, most incredibly, to black Asiatic moon bears and traumatized rescued lions. Severely abused horses let him tend their wounds. Aggressive dogs and anxious cats become puddles at his feet. Entire herds of horses, like Warwick with Cody the mustang, lie down and doze.

There's no trickery, no hypnosis, no Amazing Randi or small-town magician vanishing animals' consciousnesses like cell phones. Rather, James French, soft-spoken originator of the Trust Technique, does it through listening and by harnessing what he calls the "power of the present moment."

The first time I saw him in action in a video five years ago, I was intrigued. What was this man *doing*? What otherworldly power did he possess? I went straight to his website and purchased his video course. Could I really do this too?

I practiced it that summer. The results were astonishing. I could sit in the field, and Major, this big red being buzzing with distracted energy, would graze his way over, lie down, and fall sound asleep beside me. The key lay in a simple, meditationish practice: not thinking about the past or the future, not analyzing what was happening to or around you, rather, just *being*. Set your human agenda aside, the technique teaches, and offer your animal a peaceful "present moment." You're not saying anything. You're not asking anything. You're not *thinking* anything. You are simply holding space for them, which itself is a profound sort of listening.

"The less we think," says James in his 2023 Bologna TEDx talk, "the more connected we can become." It makes you a sort of "anchor" for the animal, he explains, prompting them to reduce their own thinking levels and attendant emotions too. Coupled with close, careful, non-judgmental listening, which James calls "mindful regard" ("Half of building trust," he says, echoing Warwick, "is that the individual feels regarded and understood"), the results are a trusting relationship and the reduction of an animal's behavioral problems.

An individual's behavior, good or bad, James says, is always driven by their perception. Their perception is driven by how they feel. Change how an animal feels by helping them connect with you in a deeply relaxed state, and their perception changes, causing those "bad" behaviors to melt away. It feels, he says, "like you're reaching out and touching their soul."

And it did.

Every time I tried it that summer, I felt a groundswell of joy and a deepening bond with my complex, challenging horse.

During our last sun-kissed days in Barbados, while Indy was busy with his little brown jobs, I decided to have another go, this time with a shy troupe of green monkeys, *Chlorocebus sabaeus*. Rangy, golden-furred Old World animals with Harlequin masks and whiplike tails, they live in the forest grove at the end of Peter's garden, assiduously avoiding all humans, descendants of those brought over on eighteenth-century slaving ships from West Africa. At the center of the grove stands a small wooden pergola. In front of the pergola sits a stone bench. And on this bench, I placed, for three days running, an non-contingent gift of three green bananas, then walked straight away.

Typically when we glimpse "exotic" wild animals, we *want* some-thing: to check a species off a "Seen It" list, for them to notice us and respond. Even a photo is a kind of taking. But the Trust Technique is about offering something—your time, your listening—without expecting anything in return.

On the fourth day, I brought my bananas and sat down inside the pergola to wait, channeling not just James, but also Cat, Con, Warwick. I slowed my breathing. I fixed my vision on one of two snails affixed to the pergola wall. I kept completely still and focused on the present moment. *No agenda*, I reminded myself; *no expectations.*

After fifteen minutes, one young male appeared, thirty feet up in a vast old banyan tree. Cautious, he circled me, coming no closer.

I carried on watching my snail.

Five minutes later, the monkey reached the ground, and I couldn't help feel a thrill of excitement. This must be why people hunt: that quick hit of adrenaline, the dash of cortisol, some oxytocin for good measure. *Settle down*, I scolded myself. *Sit still. Calm that monkey mind. Listen.* I focused on the sound of the breeze in the banyan. I wrestled my breathing into steady submission. I felt the position of each of my fingertips, each of my toes, in space.

As I did so, he approached the bench four feet away, grabbed a banana in each fist, and set upon them with a gastronome's gusto.

I mindfully regarded him from the corner of one eye.

He was a smaller monkey—a juvenile all alone—insecure—but daring. He polished off the first banana and retreated behind a log to guzzle the second. Was he hiding from me, I wondered, or his loot from other monkeys? *Stop thinking*, I thought, and with trouble did so. In four neat bites, the banana was gone. He tossed the peel and, with the third in his mouth like a dog with a favorite stick, was up and away through the banyan into the forest canopy.

My second attempt was three hours later. With no bananas left, I took some cold pasta from the fridge, a cucumber beheaded for a salad, and two children.

The first child lasted five minutes, then headed back to the penguin pool.

The second, Indy, sat quietly. We selected a snail each and stilled our minds.

After ten minutes, they materialized. A whole troupe this time: parents, babies, old folks, teens. I lost count at a dozen. They passed close by, pasta evidently insufficient to detain them. But as they departed, one monkey doubled back, a male again, bigger than the first. Indy and I focused on our breath, on the feel of our skin, on our soft snail gazing. The monkey approached the bench, raised himself onto his hind legs, and surveyed the selection on offer. He seemed unimpressed, a lunchtime latecomer to the end of the buffet. Eventually he selected one piece of penne and held it up in our direction, either in appreciation or disgust. He nibbled and was off, no hanging about or returning to refill his plate.

A minute later, a mongoose appeared at the bench. Forgetting myself, I stared straight at it. It vanished. Mongooses know the predator mind-set intimately, being voracious predators themselves.

After that, only birds seemed interested in the remains of the meal. I resolved to up the ante, buy five pounds of ripe bananas from the market, and come back this evening to await the troupe's return.

Two more visits.

No more monkeys.

Was it the wrong time of day? Did I want it too much? Had the penne put them off?

But then, at six the next morning, as Falafel and Indy trained each other in the living room, I heard a crash and a squeal from the forest. *That's them*, I thought, possibly running afoul of the collies' sense of early-morning order. I marched to the fridge, took out the chilled half watermelon we'd been keeping for breakfast, and headed down to the grove.

Hoisting it in like an offering—*Ah Zabenya*—I could tell they were watching, could feel their eyes on the back of my neck. Was this, I wondered, how wild animals feel when we spy on them, this strange, visceral feeling, only likely for them far more acute? I placed my melon reverently on the bench and retreated to the pergola.

Immediately, a young female arrived. I focused on my breathing. On the feel of my bare feet on the floor. On the birdsong in the trees. On yesterday's snails, now united and performing a slimy love scene. She hopped up onto the bench. Cautiously, she eyed me, then tucked right in.

Soon she was slurping. She gripped the half watermelon in both hands, juice dribbling joyfully down her chin. Occasionally I looked at her, practiced James's mindful regard, and went back to my snails. We danced like this for several minutes. Look at snails. Look at her. Look at snails again. And then I blew it. Head-on, predator style, I tried to line my phone up for a surreptitious snap. All that thinking. All that *wanting*. Promptly, she knocked the melon off the bench and wobbled it away into the bushes like a waiter with a heavy tray. Ashamed, I went back to breathing. The monkey went back to her meal.

At this moment, I was discovered: Indy, wondering where breakfast was hiding, appeared in the glade. I looked up. When I looked back, monkey and melon were gone.

On the last several mornings of our visit, I didn't stay down at the grove. Instead I left bananas, noncontingent, on the bench and practiced my present-momenting alone on the old stable steps. The Collie Patrol arrived to check everything was in order. The first shafts of sunlight warmed the iguana enclosure. A lambent dragonfly skimmed the pool.

And each day, the green monkeys joined me in a mango tree overhead. First one female with a babe in arms. Next two males. Then a dozen more, tails curled around branches, picking their produce like fickle market shoppers, taking a bite of green mango, tossing it to the ground, selecting another. Their glances at me became fewer and fewer. I ditched my agenda. I upped my Trust Technique. Nobody lay down to sleep. But it seemed I was now just part of the living landscape. I could scratch my nose, shift my weight, and it didn't scare them away. Perhaps they recognized, as we do apes', these shared gestures.

"Every being," James French says, "is seeking connection, even if you don't think so at the beginning."

And the monkeys weren't the only ones. A grackle came down for a birdbath beside me. *Zip-zap-zwee,* he called to another, who joined us. A gecko ran up the wall. A scaly-naped pigeon walked over my bare foot, fat and unabashed, with his purple plumage and his stubby red toes.

When I was seventeen, I spent the summer backpacking with a friend around Greece. We slept on beaches and park benches. We tromped around ruins. I read dog-eared copies of *The Iliad* and *The Odyssey.* We were invited for meals in shepherds' huts with their stoic donkeys and flocks of leggy sheep.

One early morning on the slopes of Mount Parnassus, with no money for the entrance fee, we hopped the fence at the Temple of

Apollo. And there I stood, amid the shriek of cicadas, alone in its ancient remains. There was once a column here, I knew, inscribed with three of the Delphic maxims. "Know thyself" was the only one I could remember.

I was certain I did.

But a funny thing happens as we get older. The more we learn, it seems, the less certain this certainty becomes. We read and we study and try to find out and, with a surfeit of information, often end up less sure than we were to begin with. Our ideas become stories, and our stories crystallize into self-limiting beliefs that draw us further away from those two Delphic words of wisdom.

Stories have been with us a long time. They are how we make sense of our known and unknown worlds. Bernie Krause believes they originated in the sounds of the forest. Cat Hobaiter, she told me, has a "little side gig" studying the effects of firelight on storytelling. Herb Terrace thinks language derived, in *Homo erectus*, from a need to tell stories about the location of a juicy dead mammoth.

Stories, to humans, are life. They give us hope. Religion. Shared histories on which to build stronger relationships. They're dependable, always there. Whenever I'm sad or uncertain, I reach for the same ragged literary comfort blankets: *The Corfu Trilogy*, *In Patagonia*, *Three Men in a Boat*. Stories are the pinnacle of humanity's capacity for beauty, a celebration of commonality and discovery of difference bound up into one.

But stories are dangerous too. They can make us cling to beliefs instead of seeing what's really there. They can prove our siren song, the antithesis of knowing thyself. Stories create labels and prejudices. Stories cause wars. Such is the power of stories that oppressive regimes the world over have burned books to try to stamp them out. Zipping through time and space, stories are the ultimate subjectivity, the very opposite of the present moment.

We tell stories about species. Coyotes lure dogs to their deaths (they don't). Mules are stubborn (they're not). Pit bulls are dangerous (they aren't). We tell stories about one another. And we tell stories about ourselves and our animals: My cat doesn't like me. My dog doesn't listen. I am hopeless at this. I should probably just give up. We catastrophize in such vivid detail that it affects, in an instant, our entire body chemistry.

Sobbing amid the worst of February's ice, I'd already killed, buried, and mourned the horses before I'd had my morning's first coffee.

This misty morning, though, six thousand years from ancient Delphi and twelve hundred miles from the Great Nova Scotian Glacier, I'm at ClearWind Farm, near Chapel Hill, North Carolina, to watch Nahshon Cook, a storyteller himself, at work. Nahshon is a published poet, a devotee of Mary Oliver and Toni Morrison, and as a young Black horseman in an overwhelmingly old white industry, he faces the challenges stories present daily. This makes him especially astute at recognizing their ill effects on others.

Nahshon got in late the night before from Colorado and, bundled up in a Carhartt jacket, boots, and hoodie, is nursing a head cold. His audience assembles in the farm's indoor riding arena. People mill and talk horses. I listen in. In my experience, it's an opinionated world where casual comments can erode a rider's self-belief. "You're too soft." "You should be braver." "You'll never make progress like that." Here, though, the talk seems kinder.

Nahshon's first clinic participant is a lady named Sue, riding Dexter, a chunky palomino who won't canter. The problem started, she says, when she sent Dexter to a trainer who tried and failed to teach him. Afterward, she discovered Dexter had a broken bone that made it impossible for him to do what she and the trainer had asked. "Now," she explains, "he's totally healed, but still won't canter, and I can't free myself from that guilt."

Nahshon is not your usual equestrian clinician. He doesn't

teach technique in the traditional sense. He does not scream, "Back straight!" or "Heels down!" even once. He doesn't even mention quadrants or conditioning. He works with subtler, less quantifiable stuff: with feelings, sensations, visualization, and embodiment. He helps Sue "come into her body" with a guided meditation. Imagine, he says, a safe, heavy feeling, weighing itself on Dexter from front to back. Breathe, he entreats her. Her guilty feelings, he says, are shutting Dexter down. "Horses are real embodiments of emotion, and stuck emotions stop movement. Try to unlearn being afraid of asking a clear question. Do away with that unhelpful story. Stay in your body, in this moment."

Outside, the morning mist begins to clear.

"Now, ask him to canter."

Dexter and Sue canter effortlessly around the arena.

"You did that," Nahshon tells her. "That was all you. You came in sounding like a prayer flag fluttering in the wind. But now your voice is strong."

Sue beams. The audience beams. Even Dexter seems to beam. Nahshon might not teach theory, but everyone here, audience included, has just been positively reinforced.

The clinic's off to a good start.

Throughout the day, the unhelpful influence of stories surfaces time and again.

"Communicate based on the surrender of your ideas," he tells the owner of a tense, "explosive" Arabian horse. "That 'Arabians are this or that.' That 'this horse is this or that.' Instead create a poem in this moment, a haiku."

The Arabian settles and drifts off into a doze.

"Oftentimes," Nahshon says, "we wear the inheritance of a story like hand-me-downs. And that story isn't even ours."

Later comes Mimi, a scowling, plaid-shirted lady with a

sweet-looking chestnut mare. She enters like a battleship: shoulders tight, chest puffed out. Had she hackles, they'd be raised. *This woman is scary*, I think, *and not very nice*. They frog-march a lap of the arena.

When they stop, the mare steps in close to her, and Mimi corrects her with a fierce snap of her lead rope. The audience gasps.

"Hello," says Nahshon. "How can I help you?"

This horse, Mimi says, does not respect her personal space.

"Is that important to you? That space?"

Mimi frowns. The horse, she adds, once tried to kick her, but she sent her to a trainer, and things have been OK since.

"So why is the kicking story still valuable?"

She frowns again. "Because she's rude. She gets up into my space. And I don't want her there."

"Why not?"

"Because…because she should respect me."

"Why?"

Mimi blinks. Opens her mouth. Shuts it again.

"Do you feel safe?" Nahshon asks.

Mimi struggles to find an answer. She bites her lip, a chink in her armor. "Maybe not," she admits.

For the next twenty minutes, Nahshon has her walk the arena, visualizing warmth.

"Feel safe in your body," he encourages. "Let warmth and light fill you. Listen to the sound of your mare's feet in the sand."

As he talks, she softens. Her horse softens too. They come to a stop.

"That's much better," Nahshon says. "See? Look at her. So much better. She's not going to kick you, is she?"

Mimi shakes her head.

"I want to be clear. I have no expectations of you. I don't know you." He pauses. "But I do."

Mimi's bottom lip trembles.

"Do you feel safe?"

"Yes," she says.

With this, Mimi's little mare turns to nuzzle her shoulder: a gesture so sweet, so profound, I feel tears welling. I look around. Several others are dabbing their eyes with their coat sleeves too. I sense this whole arena full of people all holding space: a collective present moment.

Mimi's battleship shoulders begin to shake with sobs.

"The question is," Nahshon continues, "are you willing to give up the story? You're not protecting yourself from this horse," he says. "You're not afraid of her. She's just a convenient substitute. A stand-in for someone else."

She's crying openly now. Her mare is soft, gently walking beside her. "I feel so embarrassed," she says. "What should I do?"

"I think you should find someone to talk to."

She nods. "I will."

"Thank you," someone calls to her from the audience.

"I've been there," calls someone else.

I feel remorse for the story I constructed so readily about her.

"There are stories we hold so as not to move forward," Nahshon says as Mimi and her mare leave the arena. "Either the story you tell is a roadblock you want to get past, or you're tricking yourself into believing you want to move on, and the story is an excuse to stop you. OK, so fucking tell us the story if you have to, and then use it to plant a flower and move on. Or don't. But don't expect me to support you being stuck in the story."

"I feel you're talking to my soul," says a lady in the audience.

"How'd you get so wise so young?" asks another.

Nahshon shrugs. "Horses feel better when they're not bogged down by the story. And so do people."

I consider the stories my own horses carry.

The average equine, unlike a cat or dog, goes through seven

homes in its lifetime. Many owners, therefore, know little about their horse's history, save for what they can guess from its behavior. Sellers tell you what you want to hear and not what they don't want you to know. Jokes abound about equine sales pitches. "Adapts well to new environments" means "has been sold fifteen times." "Sweet personality" is code for "utterly untalented." "Extremely agile" means "best of luck catching him," and "Ideal for beginners" refers to the only people inexperienced enough to consider buying him in the first place.

But with Major, I once did a little digging. His first owner, I discovered, was a lady who killed her husband and hid his body after enduring many years of unimaginable domestic abuse. Major, I learned, was her escape from the horrors of her everyday life. When she eventually went to prison, a rural farrier bought him from her. "I'll tell you something straight up," he told me when I tracked him down. "That bastard bucked me off hard twice, and I rode broncs professionally. He was kind of spoiled. So I rode the frigging hair off him. But he was a bit of a dick. He always had that in him."

What's the value, I wonder, of these stories? Do they explain my big red horse? Do I use them as an excuse? Would they have scared me off if I'd known them in advance?

Before I decided to keep Major, I spent a long time just hanging out, getting to know him, learning his quirks and idiosyncrasies, his likes and dislikes.

"Oh," his then owner exclaimed, astonished, when she watched me ride him. "He's *listening* to you."

That's because, I thought, *I've been listening to him too.*

"Being able to listen," Nahshon tells me once the clinic wraps up for the day, "requires you to be quite secure in yourself, because to listen you have to be open to fully experience what someone is sharing with you. The good *and* the bad."

He gives an example of a lady he helped recently. "'I only like hearing my animals when they tell me they love me,' she told me. That's where all her problems lay."

Horses, he says, are so exquisitely attuned to human energy that if he asks an owner to visualize their horse softening or lightening in a certain bodily area, the horse will immediately do so. I've experienced this myself; full of joy on a woodland ride, I might just think about a nice, bracing canter, and Major responds as if reading my mind. This is born out in science. A 2009 study describes how when riders were told an umbrella would open suddenly as they rode by, both their own and their mounts' heart rates rose—though the umbrella was never actually deployed.

"So what they really need from us is just to be honest. But we're taught not to trust ourselves. To play the character in some story. Being who I am in this country, I look at many things through the lens of one whose ancestors were slaves. There's still so much happening that is born out of Black people learning how not to be told who they are, but to define that for themselves. Colonialism is like dominating an animal. 'I am taking away the best of you,' it says, 'without your permission and for my benefit. I don't care how it makes you feel. I'm not listening. Don't object.'"

Nahshon tell me the meditation he's been mulling as he goes about his day.

"I'm finding that the disease of being addicted to the story empowers the problem. Empowered problems are hard to solve."

He repeats it.

"So fuck the story you tell yourself about your animal." He sips tea for his cold. "Fuck the story you tell yourself about yourself."

Back home at the farm, a torrential week makes rivers of ditches and bogs of fallow fields. March's geophony comprises the burps of saturated land. Constance and Agnes hover at the mouth of their pigloo: two old ladies at a bus stop. Siegfried rails at the skies.

But I am dry inside another indoor riding arena where, on a whim, I am taking a lesson in French dressage. Dressage, or "horse dancing" as Indy calls it, is not my usual cup of tea. It's refined. Elegant. The very opposite of my slapdash, crash-through-the-undergrowth riding "style." But to be a good dressage pair, your horse must listen closely to your minutest of cues. To be an excellent one, I suspect, you must be listening just as closely in return.

Today, though, the drum of rain on the roof of this plastic-domed structure, like an enormous polytunnel greenhouse, makes it hard to hear anything. It's akin to being beneath a firing-up 747 without ear protection: a disorienting, unremitting roar.

Six horses are in here, housed in small stalls. "Don't worry," avers my teacher. "They're used to it."

I smile back and say nothing.

Trigger, the pony I'm here to ride, is a small, stoic Haflinger, a Swiss-German horse bred for hauling logs out of the dense Black Forest. He is tough and heavy and, as a pony who gives multiple lessons each week, an eminently sensible fellow.

Yet I already feel rattled. I mount with trepidation, like an accident waiting to happen. My stomach tumbles as if I'm about to leap out of a plane. What's wrong with me? Why am I feeling this way? My teacher calls me on, but Trigger refuses to move, then shies sideways. Every step forward feels filled with fear.

"Are you OK?" she calls.

"Yes," I yell back above the din. "It's just…" Just what? That after more than forty years on horseback, I'm afraid of riding a *pony* at a *walk*? I immediately turn inward to stories. I've lost my nerve. I've forgotten, overnight, how to ride. Perhaps I never could. I imagine what my instructor must think of me, what she'll say to mutual acquaintances. "Oh, Amelia? She couldn't even ride *Trigger*."

I grit my teeth. I shove down the fear seeping into my chest. *Don't be stupid*, I tell myself. *Concentrate*. And on I go, through the most excruciating, most anxiety-racked experience of my entire horse-riding career.

My technique is abysmal. Nothing works. Trigger won't go anywhere near the far end of the arena. "*Make* him," the instructor commands, cross now. I don't. I see her sigh despite the roar of rain.

And then, halfway through the lesson, it hits me. I felt this way *before* I mounted. It's not me—*it's them*. I look back across the arena at the five stalled horses, then down at Trigger. I'm feeling *their* feelings, like a radio picking up scratchy, worried SOSs. I can feel their collective concern. Trigger's is vibrating through me. Like that windy night when Andvari and Major went a little loopy, they need to hear, to move, and they can't. If I, a human, had been in here, in this din all night, I wouldn't get used to it. I would wind up, tighter and tighter. How much worse it must be for a horse.

At last, with neither success nor calamity, the lesson ends. "Well done," my teacher says, though she knows I know she doesn't mean it.

I make a grateful escape and sit in my car, furious with myself. Haven't I learned *anything* from James, from Nahshon? I should not have ridden poor Trigger. I should have stayed in the present moment, listened to my feelings and his and not to the stories in my head. I should have said, "No thank you, not today." I shouldn't have cared what this teacher thought.

Know thyself? I laugh. I barely know anything.

I make an angry promise behind the wheel of the stationary car.

"I promise," I vow out loud, "I'll *always* listen in future. Even if it makes things uncomfortable." Then I use it to plant a flower and drive away from that rainy-day tale.

Feelings are contagious. Waves of fear or euphoria can sweep through a crowd. We feel our loved ones' tension like an icy blast when they walk in after a terrible day. "What's wrong, my little rain cloud?" I've often asked one of my children when I sense something's the matter. Inevitably, there is.

A *Scientific American* article explains that when people talk or share an experience, their brain waves synchronize, neurons firing at the same time in the same areas of the brain. Couples, researchers found, share more brain synchrony with each other than with others. "'When we're talking to each other,'" the article quotes neuroscientist Thalia Wheatley, "'we kind of create a single überbrain that isn't reducible to the sum of its parts.'" Bats and mice, the article explains, entrain brain waves too.

I remember something Nahshon said at the North Carolina clinic. "One plus one equals one if it's relationship. One plus one equals two if it's war."

The same goes for humans' relationships with animals. Alone at night at the farmhouse, an anxious thought pops into my head, and instantly Puff and Dolly wake up and start barking into the dark. I read a 2019 study examining "emotional contagion" that found heart-rate variability—the spaces between heartbeats—correlates between dog and owner, the correlation becoming stronger with duration of ownership. The Californian HeartMath Institute observed similar results between a fifteen-year-old boy named Josh and his dog, Mabel.

Still angry with myself for having failed Trigger, I make a call to Dr. Susan Fay. A doctor of psychology with a background in environmental science, Dr. Suzie helps people having trouble with their horses. But like Warwick and Nahshon, she's as much people trainer as animal trainer. "It's funny," she tells me. "I'm essentially a scientist, but I'm always asking: How can we get out of that scientific mind and become more natural in how we communicate with another being?"

What, I ask her, apart from leaning into the present moment and away from the trappings of story, is the answer?

"The breath," she tells me instantly. "And brain waves."

How you breathe, she says, can change your life. "It's the most important, actionable thing. You can reset your whole physiology in just a few seconds by changing the way you breathe. I put myself into a pattern of breath, and everything quiets. I don't do traditional meditation, but just with my breath, I quiet my thinking mind, and I'm back here in my body."

Dr. Suzie teaches a full-breath course, but I've found the "box breathing" favored by Navy SEALs (breathe in for a count of four, hold for four, out for four, hold for four, visualizing each action as one side of a square) or "birthday-cake breathing" (breathe in as if smelling lovely flowers; breathe out as if blowing out birthday-cake candles) are both quick, calming deep-breathing methods.

"From this place of quiet, calm presence," she says, "you can then create a *feeling* in the body: like a radio station that the animal can tune in to. Everything is a frequency. For example, my voice, talking to you now, is a wave that hits you, is interpreted, and then you send a sound wave back. The same exact thing happens when I think 'I would like to feel softly confident.' Now my brain waves, my physiology, align with that. My breath slows down. I become present and grounded. My body's energy signature, its frequency, has changed. It is communicating soft confidence to anybody who's listening. The strongest wave—or feeling—in the field will cause all others to entrain to it. That's the clarity with which I go to that energy. There's no static in it at all."

Sometimes, she says, "people have a hard time with what I teach, because it's about taking accountability for what we're thinking, feeling, doing. Dogs forgive us for our issues: 'You're a jerk,' they'll say, 'but I love you anyway.' Whereas horses tend to say, 'You're a jerk. Why don't you try a little harder, and when you do, then I'll change.' That confrontation of our own imperfections is hard for people. It can be very sobering and discouraging. When you decide to deal with your communication issues, you have to say to yourself, 'I'm going to put my big girl panties on and deal with whatever shows up.'"

I remember the roaring arena with Trigger. How do you know when you listen, I ask, whose issues are whose?

"First, breathe. Then perform a body scan away from the horse."

I know, from years of Thursday-night yoga, how a body scan works: a quick reading of how each part of you feels, from toes to crown, noticing without judgment all your bodily sensations.

"If you walk up to an animal and immediately feel different, you can assume that that feeling's from them, not you. I am constantly body-scanning as I work. That helps you become really solid inside: to develop a sort of emotional core strength."

Dr. Suzie compares the feeling to an inflatable punch-bag Bobo doll. "They have this shit-eating grin on their face, and no matter who or what knocks them, they bob right back up to center. That's congruence. 'What's mine is mine,' you say, 'and what's yours is yours.' We can listen, but we don't have to take it on."

Next time you're uncertain about how to listen to your animal, she advises, try it out. Harness your breathing. Do a quick body scan. Turn your internal "radio" to a positive emotion—love, warmth, companionship—to which your animal can entrain. Then be quiet, open, and curious to see what happens when you take it from there.

"No distraction. No static. No doubt," says Dr. Suzie. "And just hold it."

An end to winter is finally in sight. Daylight is a bubble-gum bubble, expanding slowly. On frostless mornings, Fräulein and Hildegard scratch about in the orchard for last year's vinegary windfalls. I spot Dolores at the edge of the wood. Sven the squirrel is out and about.

On March 24, a flock of dark-eyed juncos arrives at the farm: the year's first returnees. They settle on the terrace to refuel from their long-haul flight. I toss out seed. Feeding weary travelers is a tried-and-tested means of gaining trust between species, like Tolkien's Beorn, who shelters Bilbo and the dwarves on their way to the Lonely Mountain.

While they eat, I tie back the naked raspberry canes and discover in their midst an empty hummingbirds' nest, made of lichen and wisps of leaf. Examining this miniature marvel, I start fretting about Mervyn. Where is he? Is he OK? Will he make it back alive?

I default to Thoreau.

"I am not where my body is—I am out of my senses… What business have I in the woods, if I am thinking of something out of the woods?" he wrote in his essay *Walking* in 1862. An early version, I consider, of James French's present moment and Nahshon's "Fuck the story."

The sun is out. The juncos peck seed. I take a big birthday-cake breath.

8

"Happy Them Up"

How to Listen Through Humor, Fun, and the Biology of Joy

> "Through primrose tufts, in that green bower,
> The periwinkle trailed its wreaths,
> And 'tis my faith that every flower
> Enjoys the air it breathes."
>
> **—WILLIAM WORDSWORTH,**
> **"Lines Written in Early Spring"**
>
> "All animals, except man, know that the principal business of life is to enjoy it."
>
> **—SAMUEL BUTLER**

APRIL

April showers bring May flowers. I hold this thought as the skies lower, rain sweeps in over the mountain, and the heavens empty.

Major takes cover as the land is hammered by five-cent raindrops. Puff refuses to go out altogether and has to be carried. Andvari alone relishes the weather. Muddy from recent rolling, he stands outside as the once glacier becomes a whitewater rapid beneath him. I consider squirting on shampoo and letting the rain do the rest.

The meadow responds to the wet, warming earth with greater ebullience. Four, five, six feet…grass towers above my head, terrific for insects and mice and amphibians, but less so for equine waistlines.

Icelandics especially have the nutritional needs of an air plant. I watch the slight wobble of Andvari's bottom. The dread specter of laminitis looms. Too much sugary grass can cause horses irreparable, even fatal, hoof damage.

What to do, then, with all this grass? Something, Gal requests, that "sparks joy." So scything's out, as is getting a tractor stuck in the boggy bits. But I know exactly where joy is to be found in early April in the Northern Hemisphere. I proceed straight to my friends' farm, where their annual crop of adorable, voluble kids has just been born.

Play comes early to humans. Babies laugh at funny faces at just twelve weeks old and by eight months can make grown-ups laugh intentionally. But to baby goats, it comes even sooner. Joy incarnate, there's no woe, growls Dan, a gruff ex-military man who served years in Kandahar, that a newborn kid can't fix. Tenderly, he plucks two-day-old Hermione from her straw bed.

"Ma!" she cries, her voice weirdly human. "Ma!"

He cuddles her. She stops wailing. Brother Harry and cousin Hamish race by on wobbly legs. "It don't matter that they're going nowhere." Dan sets down Hermione to join them. "They've just gotta get there fast."

To get involved in their game, I have only to sit on the ground. Within moments, I'm part of the architecture of play. I am climbed on, nibbled, headbutted with adorable horn buds. I become a mountain to summit, a rock to race around. And all at top speed, until suddenly, temporarily, I'm a convenient pillow.

Goats and humans are a match made in heaven. There are yoga goats and therapy goats. A 2019 study found goats can distinguish between happy and unhappy human vocalizations. I have a friend whose goats hike forest trails with their Airbnb visitors. The goats have fun. The visitors have fun. Passersby have fun. Her goats are good-time girls.

Goats also possess the iron constitutions of competitive eaters. Thistles, nettles, brambles: Nothing deters them. One memorable seventh grade outing, a goat ate my geography homework. They'd have a field day with all my grass. The trouble, however, is that goats have the escapist prowess of Harry Houdini, combined with the exploratory urges of Dr. Livingstone. I look at my friends' jail-level fencing, through which these brand-new kids are currently squeezing, and think of our neighbors' tender peach trees and vegetable gardens. I'm not certain my own constitution is up to the challenge.

Who else then? A couple of cows perhaps? Our farm was once home to a dairy herd. There's still an ancient milking parlor and a wooden cross beneath which I imagine old Daisy or Clover lies at rest. Cows, I'm certain, experience joy too. For several years, we lived in southern Goa, on India's emerald West Coast, where locals, expats, and tourists gathered daily on the beach to watch the sun slip into the Indian Ocean. And every evening, the village's free-roaming holy cows, along with a scruffy selection of street dogs, gathered also. They settled on the sand, not begging for scraps or attention, simply, it seemed, relishing the shared experience of the sunset.

Animals, said Darwin, possess a sense of beauty. "There is also good reason to suspect," he went on, "that they love novelty, for its own sake." I firmly believe this is true. I imagine a pair of miniature Highland cows, all auburn bangs and curvaceous horns, watching sunsets with me over the river oxbow. (Would their moos, I wonder, retain any Scots lilt? It's unlikely. Twenty years ago, a British phonetics professor found cows moo in regional accents.)

For advice, I phone Ronald Rongen in Belgium. Veterinarian, geneticist, animal behaviorist, and the world's preeminent "cowmunicator," Ronald trains farmers, animal rescues, and even slaughterhouses to solve bovine behavioral problems. "Don't have the little ones," he shatters my daydream. "The bigger beef breeds are much more placid. What's calmer?" he adds. "An Irish wolfhound or a Chihuahua?"

Ronald defines joy in a cow as contentment. And contentment, he says, is easy to achieve once their basic needs—food, water, company— are met and so long as you know how to cowmunicate with them. "Never act like a predator. A predator will separate one cow from the herd, then kill it. So always provide company. And if you approach a cow directly from the front, from their perspective, it's always an attack. That means it's a suicide mission for us. When I approach a cow or bull or herd, it's always from the side, typically from the left-hand side. The left eye triggers the right brain, which contains the 'Is it safe or not?' processing. If I approach a cow from the right-hand side, she is extra nervous, confused, suspicious, because the left eye can't check the situation. After that, if you run into a problem, simply look at it through their eyes. Both of them.

"Still," he repeats, "the bigger the cow, the calmer, especially for a beginner cow keeper. Within every breed, there are exceptions. But in general, Dexters and minicows—forget it. And Dahomey dwarf cattle? They fly like pigeons. But Herefords, Belgian Blues, if well treated and understood, they'll be calm and happy."

Yet there's a disadvantage. A pony produces about thirty pounds of manure a day, a horse fifty, but a beef-bred cow emits roughly one hundred, much of it in jet streams of liquid manure. Herefords may be joyful, but they're also full of another substance altogether. For excremental reasons, I rule them out.

Then I remember. In a farmyard down the road lives an animal who sparks joy for my husband every time we pass by. "There's the creature!" he cries. "He's eating! He's sleeping! He's playing with his Jolly Ball!"

The *creature* is a Jacob sheep, a small, mythical-looking ram sporting three curlicue horns. The Jacob is a multihorned or "polycerate" breed. They can possess six or even eight apiece. They are spotted black-and-white and considered "unimproved": a compliment, apparently, when it comes to an ovine. Of Syrian ancestry, the Jacob was particularly popular with novelty-loving Victorian gentry, who grazed them between their follies and their hahas. I imagine a flock of little woolly triceratopses trimming our meadow grass efficiently, sparking joy in all the walkers who traipse the rail trail behind the farm.

When my grandma was a little girl, each spring she received a pet black lamb for bottle feeding. Some years, Blackie was cuddly; others, skittish. Sometimes he was an explorer, ranging off to poke about in the pantry.

Sixty years later, we would sit by the fire, and she'd tell me stories of Blackie, who would disappear "back to his flock" each summer to reappear as Sunday lunch with mint sauce on the side. I watched sheep on vertiginous Welsh mountainsides. My great-uncle convinced me that these hardy individuals had two legs shorter than the others and that if they changed direction, they'd fall over. I also watched them being treated quite roughly. Doused in sheep dip. Sheared peremptorily. Butchered, like Blackie, at six months old.

To call someone a sheep is an insult; it smacks of being a blind follower. Aristotle held a low opinion of sheep. "Of all quadrupeds," he said, "it is the most foolish." But I always thought sheep cleverer than people gave them credit for. Flocks of starlings perform a murmuration, schools of dolphins hunt together, and we marvel at it. Sheep escape and make undulating patterns while blocking a country lane, and everyone calls them stupid. I never thought they were stupid. For a start, they rarely wanted to hang about around humans, which seemed, under the circumstances, eminently sensible to me.

"I bet your ancestors knew at least a few of their sheep by name," said cowmunicator Ronald, "and even years later could remember the characters of certain individuals—for example, that one old ewe with a certain black spot on her back. But now farm animals are commodities, not individuals."

I know nothing, it occurs to me, about the world's millions of she-ople. Are they as happy as pigs or pet chickens? (I've found few things more mutually heartwarming than sitting on a porch in the sun with a chicken asleep on my lap.) Lambs are joyful little Zebedees, with four springs for feet. But does this winsome personality slough away by adulthood (which, like Blackie, few sheep achieve)?

Returning briefly to science tells me a thing or two. Like us, sheep experience emotional contagion (including contagious yawning) and exhibit "cognitive bias": Positive experiences make them optimistic and negative experiences, pessimistic. Sheep have strong social bonds and form "preferential relationships"—called friendships when we're talking about people. Lambs love being petted. Sheep recognize human faces, even in two-dimensional picture form.

But when I google "sheep joy" and "sheep expert," some unsavory results come up. Number one: "Unrivaled Sheep Insemination Instrument."

Then I find Barbara O'Brien.

Barbara is an award-winning commercial animal photographer and supplier of animal actors for commercials, TV, and movies. Over the last three decades, she's worked on set with African elephants, Siberian tigers, camels, a black jaguar, and "a polar bear from New Jersey," along with most species of farm animal and hundreds, perhaps thousands, of domestic dogs and cats.

"No matter the animal," she tells me, "we have the same motto on set: 'Best Day Ever!' The key for performing animals is knowing what brings them joy, then making sure to 'happy them up.' The moment my performer's not having fun, that's it, it's unequivocal. The day is done."

After work, though, Barbara returns home to her forty-acre Wisconsin farm, where her favorite species awaits her.

"Their names," she tells me, "are Lady Margaret, Beatrice, Gary Cooper, Jimmy Stewart, Humphrey Bogart, Jean Arthur, and Ingrid Bergman."

All seven are pet sheep.

"Humphrey Bogart's this little toughie, whereas Gary Cooper's kind of laconic. Jimmy Stewart just wants to be loved, and Ingrid Bergman's a small Icelandic sheep, the only one with horns, and has an 'I could knock you all over' air to her, just the way I imagine Ingrid would've. Jean Arthur's a fun girl, and Beatrice is the oldest, a Border Leicester-Cheviot sheep who is completely enamored with human beings—especially children. Lady Margaret's a Suffolk meat sheep, a big black-and-white lady, over one hundred pounds, and very much the boss."

Humans, Barbara says, underestimate sheep personalities.

"They have no jobs except to give us wool or meat, so we just leave them in a field. Then we think of them as fearful animals. But the reason is *exposure*. What have they been around? I'm sure a sheep who lives in a petting zoo is pretty nonchalant about everything."

A sheep, she says, would be a wonderful emotional support animal if exposed to the world.

"They are very much individuals, as much as any dog or cat. They might move as a flock, but that's because they have few natural defenses except for safety in numbers."

"How," I ask, "do her sheep express happiness?"

"Even the adults," she says, "play. The wethers especially will butt heads, and Ingrid does that sometimes too."

But if a human wants to share joy with a sheep, she adds, it's about relaxing, not playing.

"After work, I sit down with my back against the barn, and they all sniff me up and down. Where've I been? What've I been doing? Then they come in close, and each tells me with body language where they'd like to be scratched: behind their ears, under their chests and armpits. If you're soft and quiet, they vie for position. Their ears get floppy. Their gaze softens. I bet our heartbeats synchronize. When a sheep wants to be your friend, it's really gratifying. Because here is a creature who's afraid of humans shoving them, pushing them, making them go somewhere they don't want to go. But if you have quiet energy and prove to them over time that you're the source of all goodness, magic happens."

Goodness, Barbara says, comes in many forms, including plates of spaghetti and blueberry pies. Or even in the form of training. Beanie, I learn, a Valais sheep in Australia, holds the Guinness World Record for most tricks performed by a sheep in a minute. Beanie performed eleven but knows, in total, over one hundred."

It all sounds great so far.

But there's a make-or-break I need to know. Outside the farm, another storm blusters. Thunder cracks. Puff watches me balefully. The pigs do their best impression of a German weather house. Major flinches at raindrops.

"What," I ask Barbara, "do sheep think about the weather?"

I imagine myself negotiating with yet another species to *please* just step outside.

"Oh, they don't mind rain at all," Barbara replies. "And they quite enjoy snow."

I breathe a sigh of relief.

"It's wind they don't like. On windy days, Lady Margaret will decide they're all staying indoors, and indoors they stay."

Highly unstupid behavior, I decide, that I can definitely live with.

A fortnight later, on a brief visit to family in England's Peak District, I get better acquainted with some ovine individuals. The farm on which we're staying is home to a small flock of Dutch Zwartbles sheep along with several Border Leicesters, like Barbara's Beatrice: a big, adorable breed with permanently surprised, flappy antenna ears.

I rise early to find sheep grazing the lawn outside the kitchen door. When I open it, their heads swivel, eyes wide, like sheep in an Aardman animation.

I recall some advice Barbara gave me: "If you visit a strange flock and you're quiet, without an 'I'm going to make you do something' type energy, then their curiosity will win them over. Get small. Get quiet. And just wait."

I sit down on the doormat. I hunch over a bit.

There's a pause, then they come closer and, one by one, plop down, Trust Technique-ish, for a rest. I don't know if these hardy Northerners enjoy a sunset as much as a Goan cow, but they seem partial to a sunrise.

Gal comes outside to join us. We sit. They sit. Together we spark joy. The sun comes up.

"Almost all creativity," eminent twentieth-century psychologist Abraham Maslow is said to have said, "involves purposeful play."

Though Maslow might never have known it, this statement, I discover one rainy April morning, is true of both scientists and bumblebees. Because in 2022, I read, these two disparate species together created the most joyful study ever undertaken: "Do Bumble Bees Play?" Yes, came the answer, to which the most adorable footage of *Bombus terrestris* playing with wooden balls attests.

Younger bees play more, says zoologist Lars Chittka in an online lecture I attend on the subject, and males more than females. "Males," says Lars, "do absolutely no useful work for the colony so have quite a bit more time on their hands."

Ain't that the truth, I think as I hear my teenagers kicking a soccer ball around the workshop.

So why do they do it? Because, suggests Chittka, it's *fun*.

Fun builds bonds between species. You play with your dog. Your dog tries to play with your cat. For decades, Denise Herzing has played with generations of wild Bahaman dolphins. And in 1998, neuroscientist Jaak Panksepp famously had fun upon discovering that rats like to be tickled by humans.

One morning, the story goes, Panksepp announced to an assistant, "Let's go tickle some rats!" To see what the rats would have to say, a graduate student adapted a device for recording bats' ultrasonic communication. The results were heartwarming: The rats loved it, and they *laughed*. In a 1998 interview with the *Toledo Blade*, Panksepp said, "Lo and behold, it sounded like a playground!" The resultant search for the neural circuits responsible for laughter he called, winningly, the "biology of joy."

Science, Dr. Jenifer Zeligs told me, is learning for the sake of finding out; training is doing something with those findings. But Jaak Panksepp and his rats accomplished both. He listened, learned, and with that built a mutually rewarding relationship. Panksepp trained his rats to expect good things to happen. In return, his rats trained their human to tickle them more.

"We got totally addicted to this," Panksepp told *Discover*. "Give an animal a really good time, you know? They become so fond of you, it's unbelievable."

I relish this lovely image. A serious neuroscientist, feeling that same triumph and elation I experience when my scratches are good enough to topple Constance or Agnes sideways. And mutual joy extends into even unlikelier interspecies waters: I watch a YouTube video of conservationist and scuba diver Jim Abernethy petting a fourteen-foot wild tiger shark named Tarantino, whom he's known for more than a decade. He pets, gets caught in a current, swims back, and she returns for more. Jim uses a hand signal, rubbing his thumb to his fingertips to ask, "Would you like to be petted?" A sort of reverse-engineered "go" button. In response, Tarantino comes back for further petting.

Sharks, I read later, have friends. They have personalities. They can distinguish between humans, regardless of changes of wet suit. ("There's no such thing as shark-infested water," says shark scientist Andy Nosal. "If anything, the ocean is infested with humans.")

This reminds me of when we took our then toddlers snorkeling with rays and nurse sharks on a reef off the coast of Belize. Though nurse sharks account for the fourth most human shark bites globally, we weren't worried for our brood, each in their little yellow life jacket and clutching a neon pool noodle. These sharks had not been fed by humans. We weren't bothering them. We were simply bobbing about in the gin-clear sea, undemanding visitors admiring the company as fishy life went on as usual around us. Eagle and manta rays skimmed the soles of our feet with their smooth, flat backs. Perhaps they were testing or tasting us. But it seemed like joy and play to me.

I hear something similar in a seminar on pangolins from Gareth Thomas, a volunteer who helps rescue this diminutive species in South Africa. Pangolins, the only mammal with scales, have existed for some fifty million years and are the most trafficked creature on Earth. Complicating their preservation, however, is that they can't be fed in captivity; they must be walked to find their requisite daily two hundred thousand ants in the wild. This takes hours, and though apparently solitary by nature, they develop, Gareth says, strong bonds with their human caretakers—even, like Sedna, playing favorites.

"If you look at any animal with an incredible amount of awe and wonder and love, it's hard for them not to respond positively," he says. Within a couple of days, most rescuees accept the volunteers as their companions. "The connection," he adds, "is fostered in the present moment."

I watch a video of a pangolin playing in a pond. Even the most hardened anthropocentrist would be hard-pressed not to call it joy.

"Conditioning and society," says Gareth, "prevent us from softening into deeper connection with creatures."

Some animals move beyond play to display what we humans consider a sense of humor. Koko the gorilla reportedly had a wicked one, tying primatologist Penny Patterson's shoelaces together, then signing "Chase." Cat Hobaiter's young wild chimps attempt to goad her into playing. I connect with Aimee Morgana, animal behaviorist, who tells me her African gray parrot, N'kisi, "has a wonderful sense of humor. He laughs appropriately and sometimes is quicker to laugh at jokes than I am. He can also be snarky. One time I walked into the room and said, in passing, 'Hey, weirdo.' He replied in an arch tone, '*You're* the weirdo—look at you.'"

On another occasion, he teased Aimee's friend, primatologist Jane Goodall.

"Jane was making her famous chimpanzee 'pant hoot' for him, and N'kisi kept telling her she sounded like a cat. She'd say, 'That's not a cat, it's a chimp,' and he'd reply, 'Sounds like a cat.' He also laughs at his own jokes. He once called someone a 'featherhead,' then laughed like crazy. I'm not sure of the full meaning from an avian perspective, but he seemed to think it very funny."

Puff's favorite game involves lingering in front of me with my shoe in her mouth, then racing off as I attempt to take it, to invite a game of feinting, bluffing, catch-me-if-you-can. An identical story appears in Darwin's *Descent of Man*. "Dogs shew what may be fairly called a sense of humor," Darwin concluded, "as distinct from mere play."

But some, like Beanie the record-breaking sheep and her owner, Noeline Cassettari, harness the power of play to push the edges of human-animal communication. I meet with Pilley Bianchi, musician, author, and dog trainer, to ask about the part play played in training her family's famous border collie, Chaser, who, by her death in 2019, knew 1,022 words: not just nouns, but also verbs like *take* and *find* and prepositions like *to*.

"Everything Chaser learned," Pilley tells me from her Brooklyn home, "was not by rote, but through play. It was joyful for her. It was fun. If it wasn't, she'd just tap out, with this air of 'I'm tired of this. You like it? *You* do it.'"

Play, says Pilley, whether it's with your animals or your kids, is innately reinforcing and second to none at building your bond. "Dogs and other large species play in the wild, with big, elaborate movements that could attract predators and expend energy they might need to save. So why do they do it? Because it feels good, in the joy of the moment."

Chaser, by any measure, was an extraordinary canine. But learning through play often sees extraordinary results.

"Take Jennifer Arnold in Atlanta," says Pilley. Jennifer runs Canine Assistants, a charity training service dogs for humans in need. "She has a 90 percent success rate training service dogs by teaching through play. That's versus the typical 70 percent fail rate in teaching service dogs with traditional methods."

So how, I ask Pilley, can we best play with our dogs?

"Often we ask dogs to do things they're not so fond of, but with play, your dog is doing something they like. So pay attention to what makes them tick, and build on that. If they come find you, it's like your kid pulling on your sleeve: Don't ignore them. Ask what they want."

Usually for dogs, she says, the answer is engagement. "They rarely want to play by themselves. Chaser would go get a toy and toss it into your lap. If you did nothing, she'd get another and another until you had a whole pile of toys and could ignore her no more."

She refers me to Marc Bekoff's "golden rules of fair play" in dogs.

"Ask first. Communicate clearly. Mind your manners, and admit when you're wrong. Stop when you're still getting yeses. Don't keep going till you get nos. And always," Pilley smiles, "let the puppy win."

After a silent winter, by mid-April, the biophony is back. Yet more heavy rains have brought out the spring peepers. At night, I stand outside and drink in the sound of love: ten thousand little green frogs, *Pseudacris crucifer*, chorusing, "Pick me! Pick me!" Bit by bit, summer's visitors are returning. Geese honk in the dead of night, following the river upstream.

An Easter basket of crocuses emerges overnight outside the kitchen door, candy pink and baby blue. I gasp to see the year's first three honeybees, buzzing languorously from flower to flower. Do they find joy, I wonder, in these first tastes of summer? I can't see why not:. A 2016 study found bumblebees experience a dopamine-fueled "positive emotion-like state" when they receive an unexpected sucrose reward. Across the river, calves rampage. Mother cows run and jink after spending all winter in the barn. Andvari bites Major's bottom and gallops away. My friend's mare births a foal, and within days, he's trying to play with the big boys.

And love is, literally, in the air.

Our pigeons build a nest in the barn, using bits of Andvari's fluffy, shedding coat. Starlings perform hasty assignations in the oaks, like spring breakers on the razz in Florida. Siegfried mounts a stump and slowly revolves, a Phasianian weathervane announcing his territory. *Kah-kah!* he cries, wings akimbo, his call a truncated rooster crow. *Kah-kah!* Flap, flap. *Cluck! Cluck! Cluck!* He becomes a squeaky toy. *Cha-chaaa! Cha-chaaa!*

The bald eagles' displays are the most startling. Their courtship is a furious, screaming one. They plummet from the sky toward the river, roar upward at the last minute, tangle and squeal and plummet again, reminding me of Tennyson's "The Eagle."

> The wrinkled sea beneath him crawls;
> He watches from his mountain walls,
> And like a thunderbolt he falls.

The inevitable outcome of all this loving? Babies. And babies mean more joy, more play. The trail cam rewards us with fawns and raccoon kits. We hope the bobcat brings a kitten.

On April 15, I let the pigs out and run away. They chase me at a porcine gallop. Experimentally, I let Puff out too. She play-bows to them, races off. Agnes is game and pursues her for a few steps, then gives up and goes back to grazing. Puff tries again. She performs a cute spin, grins at them, picks up a stick, and off she goes. This time, Constance gives chase, but Puff's too fast and furious. It's like a Morris Minor chasing a Ford Mustang.

Nevertheless, their joy is carved into their bodies. In Puff's wide grin. In Agnes's wagging tail. I'm out of breath and laughing. Then I have an idea. I fetch the boys' soccer ball. Now everyone's invested in the game: Puff, Constance, Agnes, and me, a tri-species match. Puff leaps on top of it. The pigs snuffle it along. I give it a kick, and they all pursue. Then rain halts play. Puff concedes. Game over.

With the weather's general improvement comes the chance to have fun with Andvari too. We have come a long way since those dire first days over a year ago, when even touch was beyond his comfort zone. As with Major in those early months, we've hung out until my nose is numb, then soaked, then sunburnt. But he's still jumpy. Wary. One day, under saddle, I ask too much. He panics. I bail from his back and eat sand.

So I decide to pause all "serious" riding activity and try for some play instead.

I'm not sure this little Icelandic has ever had fun with people. Everything he learned before me had a function: no silly "bang-bangs" for him, a cue Indy transfers to the pop of an old-fashioned popgun, just for fun.

I sign us up for an equine trick-training group. There's no reason for any of these tricks: the bow, the kiss, the high-stepping Spanish walk (though really there is: fun builds trust, and trust equals safety, and safety, hopefully, means fewer mouthfuls of sand). I'm thrilled to find that two of its members are Beanie and Noeline, who also, with her miniature horse, Rose, holds the world record for most tricks performed by a horse in a minute.

Throughout April, both soggy and sunny, I teach Andvari, via positive reinforcement, to bow. To "side pass": walking sideways, crossing one front foot over the other. To touch a target, first stationary, then moving. To walk over a flapping tarpaulin.

At first, he doesn't understand it's a game, not a task or a test. He seems worried about getting the answer wrong. I understand this feeling: I'm British after all. "We Aim to Please" is a national motto, and there's always part of me worried about getting told off by the authorities. For him, though, *I* am the authorities—that parent with high expectations, that jobsworth official, that airport officer at whom you grin guiltily, though you tell yourself not to.

So we go in teeny-tiny steps. I cross one leg over the other and step sideways. He follows suit. I lavish praise on him. Major watches gamely and receives his own praise for being a good spectator.

It's not linear progress. Sometimes my timing is off, or I'm not that clear. Sometimes he seems to have forgotten what we played with yesterday. But these sessions are always successful, because they're fun. I laugh. I grin. I tell him he's the cleverest boy. I read a 2019 study on how horses understand both our facial expressions and our nonverbal vocalizations and do it all the more. I also read a detailed study on their own, surprisingly various facial expressions. They have, it informs me, seventeen basic ones, compared to chimps' fourteen and humans' twenty-seven. They tell me he's enjoying it too.

When frustration threatens, I remember Harvard brain scientist Jill Bolte Taylor's "ninety-second rule of emotion." Chemically, she says, our emotions last in our bodies for ninety seconds. If we're feeling them for longer, it's because we're *choosing* to stay in that loop. The reminder works a charm. Frustration wanes. Fun returns. We always end on a high note before Andvari gets bored.

And soon something astonishing starts to happen. This shy little horse begins to *instigate* our interactions. He lips my palms, as he would groom a fellow horse. He reaches out to nibble my sleeve, steps back, and watches for a reaction. I laugh. He sighs, as if relieved his joke was understood.

I buy a rainbow umbrella, invite him to touch it when it's closed, then open, then flapping wildly. He boops it experimentally. He lets Indy dance around him and swing a hay bag, lasso-style, an inch from his nose. My friend, a horse breeder, can't believe it. Even "normal" horses, she says, won't stand for that malarkey.

It's not, I'm confident, just about the rewards. They aren't impressive: simply, praise and scratches and plain old hay, all of which he gets anyway. Free from halters and lead ropes, he can leave our play sessions at any moment to munch from a bag of identical hay always hanging in the shelter.

So why does he do it? For the feel-good factor, I believe: Panksepp's "biology of joy." And perhaps due to "contra-freeloading," a term coined in 1963 by psychologist Glen Jensen to describe how many animals prefer to make an effort to obtain food rather than simply receive that food "for free." In his original study, Jensen gave two hundred rats the choice between accessing food in a bowl or the same stuff from a pedal-operated dispenser that the rat had to push a random number of times—from forty to a whopping 1,280.

The rats overwhelmingly chose the pedal.

Theories suggest that instinctual urges play a part in this phenomenon: the desire to hunt or forage, the need to locate alternative food sources as well as those readily available (both part of Panksepp's SEEKING system). Researchers investigated further and found that mice, birds, primates, wolves, gerbils, giraffes, horses, and pigeons all chose will-work-for-food over the free-buffet option.

The only species that didn't?

Cats.

To no one's surprise, ever.

How do you play with a cat?

Though Winnie will deign to chase a catnip mouse or a feather at the end of a feline fishing line, only Cairo, our third child, has succeeded in playing in person. He'll pop his head around a doorframe or up through the banister, then scramble away, and she'll race after him. She won't do so with anyone else, will bestow upon you a look of withering pity if you so much as try. Only Cairo, it seems, makes convincing prey.

Are we doing something wrong?

I Zoom with Dr. Mikel Delgado, author, research scientist, and cat behaviorist. One of her own cats, Ruby, joins us, stationed on a cat tree in the window.

To understand cats, Mikel tells me, we must consider their history of domestication. "That process has been relatively short: only eight thousand to twelve thousand years, versus thirty thousand years for dogs. In that time, we haven't made a lot of changes in them genetically, so they still possess a lot of their original behaviors. Certainly, we've made them friendlier and probably more tolerant of other cats. But other than that, they still have all the catty instincts that other feline species have."

Because of this, she says, some feel cats are only semidomesticated—the very thing many people appreciate about them. "It's that element of wildness and mystery. That we have this companion who will sit on our laps and purr, but who still has these wild instincts. Also cats are pretty choosy about who they like, so there's a feeling of being special if they pick you."

And rejection, I think of Winnie, if they don't.

Early socialization, Mikel tells me, can influence cats' behavior and make them more "doglike." "But their socialization window is *much* younger than a dog's. For dogs, it's three to sixteen weeks, and for cats, two to seven weeks. Most people get a puppy still within that window, but most cat owners don't."

"How then," I ask, "bearing all this in mind, can a person share joy with their cat?"

"Pseudo-predatory behavior," Mikel replies. "For dogs, play has a much more social role. You're tossing a ball, and they're bringing it back to you. For cats, think of yourself more as personal trainer than playmate: the Wizard of Oz behind the curtain making 'birds' come out of nowhere. Help them simulate their natural hunting behavior. Move toys in ways that resemble prey. Cats hide and stalk while hunting. Notice how your environment might help or hinder that."

Short bursts of play, she says, mimic most closely how cats normally hunt. "Kittens tend to play for longer than older cats, perhaps because older ones are better at hunting, so it takes them less time to get the job done."

As for joy, she says, recognize that for many cats, affection comes in the form of a drive-by.

"Humans want to touch and kiss and stroke and touch some more. But cats' social interactions tend to be brief, frequent, low-intensity drive-bys: a quick nose touch, a quick rub, then leave."

That doesn't mean, she says, they don't like you.

A study performed in the 1980s, Mikel tells me, showed that if you let a cat initiate an interaction, it's more positive and lasts longer. "Cats are an excellent lesson in consent and respect. Therefore," she says, "you can train some cooperative care, a feline-specific iteration of Andvari's 'go' button.

"With one of my cats, I hold out my finger to her. If she sniffs it and rubs against it, then I pet her once. If she doesn't, I don't. Then I offer it again: same thing. A lot of times, she simply turns away. Being able to offer a choice is so powerful."

But most of all, Mikel reminds me, never assume. Her new Purdue University study, for instance, found a surprising 40.9 percent of cats love to play fetch.

"Try it out. And watch closely. Otherwise, you might miss something."

Just like my big red horse, I tell her. One winter, years ago, I was walking a fence line, clearing it of fallen branches. I picked one up and tossed it across the snow, then watched in disbelief as Major trotted over to it, sniffed it, picked it up between his teeth, and brought it back.

Bewildered, I took the stick from his mouth and tossed it again.

Again, he fetched.

"Can you record this?" I asked Gal.

He obliged as I tossed, and Major retrieved, a dozen times more.

Finally, tiring of the game, Major wandered away. And I was left with the distinct impression that if we're not listening closely, we overlook interspecies invitations to play at every turn.

It's not often an eighty-six-year-old makes you feel old-fashioned. Born in 1938, legendary dog trainer Turid Rugaas lives in Kristiansund, on Norway's southwest coast, overlooking a sheltered stretch of Atlantic

Ocean. Her book *On Talking Terms with Dogs*, which describes canine "calming signals," is a permanent dog-training bestseller in many countries. Calming signals, she says, are the subtle signs dogs display—a yawn, a sniff, a turning away—both to one another and to us, to keep the peace. They may use them to show others they're not a threat, to comfort themselves if they're worried about a situation, or to show us their concern: a variation on Warwick's stress indicators.

Turid has two white pigtails and the doughty, formidable energy of the grandmother in *The Witches*. She does not mince words.

"Fifty-five years ago," she tells me, "I taught a dog to speak. As soon as I'd done it, I regretted it, and I've never done it again. But that dog actually said 'Mama.'"

She won't tell me how she did it because, she says, "people get ideas. Though it's not as hard as you might think." But she thought it a great trick until one day, she and her dog stopped at a fast-food kiosk in Oslo.

"My dog put his paws on the counter, looked straight at the man behind it, and said, 'Mama.' And that man," she sighs regretfully, "went quite green in the face and had a heart attack. So that was that: over and out."

Fortunately, she adds, he didn't die.

"But at that moment, I understood how idiotic it is to go overboard with things that are so unnatural to animals."

(Were anyone else to have told me this story, I would have giggled nervously and asked if it was metaphorical. With Turid, however, there's no way this is anything but the truth. I look it up after our call to find a surprising number of YouTube clips confirming it is possible.)

For instance, she instructs me, you should not teach your dog tricks. You probably shouldn't even teach them to sit.

"I have not told a dog to sit in twenty-five years. And I don't need it. If I want my dog to sit or lie down, I sit down myself. It takes seconds or maybe a couple of minutes for the dog to do the same. Just shut up and show calmness with your own body. That, I find, is one of the most difficult things to teach people. To sit down and shut up."

Turid's advice may be stern, but her message is all about joy.

"Many years ago," she says, "I was in Spain for a seminar. I had plenty of time before it, so I sat outside my hotel observing dogs. Out of sixty-four who walked past me, every one was walking at leisure, sniffing around happily on a loose leash. No shouting, no commanding. And I said, 'Wow, isn't Spain wonderful?' Then I entered the seminar with two hundred dog trainers—and they were all idiots. Commanding, commanding…marching about. And I said, 'Aha! So it's not the ordinary people who're at fault. It's the *trainers*.' And why do we do it? To show how good we are!"

Never, she commands me, take your dog to training. "Instead, follow this advice."

She barks out bullet points. Obedient, I scribble them down.

Watch your dogs' calming signals very carefully.

Use your own body language just as carefully.

Shut up as much as possible, unless it's a few nice words here and there.

Stop commanding. Nobody likes that.

Cultivate a connection that's not based on your dog doing tricks to show how good he is. Nobody likes that either.

Let your dog explore and use his senses.

Put yourself in your dog's position. *How would I feel? What would I like to happen?*

And *relax* a little. Be with your dog without making him do anything.

"I get all the pepper in the world from everywhere," she concludes, "because people don't understand. But I couldn't care less. I might be wrong in many things, but in this, I'm right. Dogs are so like us: similar brain, physiology, emotions. Imagine being dragged along to do silly tricks all the time. I would go crazy."

I'm suddenly much less proud of our own silly tricks. Every one of these points makes sense. But surely it's also about the individuals. Puff seems to enjoy our interspecies training sessions, Andvari too. Am I actually a drill sergeant, missing all their calming signals and stress indicators?

I decide to listen closer next time before doing away with the bows and the bang-bangs altogether.

But I don't tell Turid that.

Early in the morning on April 25, I'm feeding the horses breakfast when I hear Agnes up a tree.

I stop and frown. What on *earth*?

There she is again, squealing her early-morning *Wheeeee* from somewhere high overhead.

Another *Wheeeee* joins in—Constance's, this time, only too high-pitched—then another and another.

Soon, there are dozens of expectant piglets requesting breakfast fifty feet above me.

I start to laugh. The starlings have learned to imitate the pigs.

Wheeeeee. Wheeeee. Wheeeee.

Constance and Agnes emerge sleepily from their pigloo, seemingly as astonished as me.

When I was a little girl, suburban magpies caused householders no end of trouble on Sunday afternoons. A sudden whoop of car alarm would send them running for their keys. The *brrrring* of a rotary phone and later the chirp of a digital one would have them hastening back indoors. I assumed the magpies did it on purpose, for fun, to see the silly humans run.

We adults, however, tend to think of birdsong as eminently functional, performed by males to guard territory or attract a mate. But a 2023 study on zebra finches says males "sing prolifically also outside the breeding season, both alone and in company and mostly when with their long-term mate." Female birds, other studies have found, sing too. Couldn't part of the reason for birdsong simply be *joy*? We humans hum and *tra-la-la* constantly to express our own feelings of happiness. It's strange to think that in an ocean of organisms, we'd be alone on that musical raft—especially when accompanied by such lyrical species.

Birds, I conclude, sing for joy, just as baby goats bounce, bees play with balls, rats like tickling, cows watch sunsets, sheep enjoy company, and big red horses fetch sticks.

I am leaning on my manure fork, thinking all this, when I feel a gentle tap on my elbow. I turn to find Andvari has left his breakfast. His velveteen nose taps me again. *Want to play?* he's saying.

My heart lifts, the way it does amid a transcendent dawn chorus.

So I drop my shit shovel and play. We weave in and out of poles. We walk forward and backward and from side to side. We run flat out, then screech to a stop. Pleased with ourselves, we summit an old wooden pallet as if it were a winners' podium. We hold each of our feet in turn in the air. We spin in circles. We bow to our imaginary crowd.

It's not Olympic stuff, but having a formerly fearful animal seek you out and *want* to play is the most rewarded I've ever felt. It goes beyond dopamine and oxytocin, beyond pleasure or an ego boost or the satisfaction of a job well done. Gaining another species' trust, unleashing their joy, is something deeper. It makes me think of those therianthropes in the ancient caves of Lascaux and Sulawesi. Somehow, you have melded. You have forged a new relationship; recreation has become re-creation. You are working or playing as one. It's an animal saying, "You hear me." They hear you hearing them. And I've experienced no purer joy than that, except after giving birth to a child. But where birth is a parting of ways, a separation, playing is concord, from the Latin *con*, "together," and *cor*, "heart." You're connected; you can imagine for a moment the way a flock of starlings feels in murmuration. You shrug off your humanness. You keen. You soar.

Later I fetch the soccer ball to take out with the dogs and pigs. Major, I know, will play ball too. We once spent a rainy afternoon pushing a yoga ball back and forth. And Andvari, I'm sure, could learn. I envisage a great interspecies soccer match, with the starlings cheering us on from the sidelines, and wonder if some Jacob sheep would join in too. They already have the matching jerseys.

9

Training the Untrainable

How to Listen When Animals Say No

> "All his life he'd been good, as far as he could,
> And the poor little beast had done all that he
>   should.
> But this morning he swore, by Odin and Thor
> And the Canine Valhalla—he'd stand it no more!"
>
> —RUPERT BROOKE, "The Little Dog's Day"

MAY

The third consecutive day of sunshine makes winter a liminal memory. In the garden, tulips tilt their faces to the sun. A blazing yellow forsythia (the "Easter bush," a month behind) has kindled against the farmhouse wall. A glory of dandelions rejoices in the orchard. A Day-Glo inchworm arching across the terrace rears up, affronted, when I offer to transport him to safety, away from hungry beaks.

Birds pour back into our river valley: red-winged blackbirds, robins, waxwings. Courting ospreys squeal above the river. Bald eagles twitter: a silly sound for such a brawny bird.

Dusk doesn't silence the exuberance. Tree swallows and barn swallows swoop and *chip-chip-chip*. Nighthawks on stealth flights whoosh and peent and boom. The spring peepers crescendo in an evening rain, refusing to be outvoiced.

Beneath the frothing apple trees, Siegfried performs a flamboyant courtship dance for Fräulein and Hildegard. He shakes and shimmies sideways, puffed up like a fat flamenco dancer. *No*, the ladies answer. He perseveres, a self-assured gigolo. Sven the squirrel busies himself in his cherry tree. Young chipmunks leap through the lilacs: sweet raisin-eyed things, their faces framed by two white stripes. The more I present-moment on the doorstep, the less they chitter their appalled "Get lost!"s.

Baby starlings have hatched in the barn. An exigent chatter arises as parents approach, beaks stuffed with slugs and spiders, oaken orgies long forgotten, and wanes when they fly off for more. I empathize. For years, I churned out twenty-one individual meals, plus snacks, per day. The pigeons have nestlings too, though not all survive. I find a chick dead on the ground, naked save for its fuzzy bottom, and one cracked egg the color of maritime fog.

We, like female pheasants and Mother Nature, say no a lot in our interactions with animals: Stop it, drop it, don't do that. At other times, we ignore them completely.

On Friday, late for an appointment, I tear through the kitchen and swipe away the wet nose that touches my palm in a bid for connection.

"Sorry, Dolly," I apologize. "Good girl."

We don't like it, of course, when the roles reverse.

During one memorable ride—or lack of it—back in 2019, Major planted his feet in a driveway and refused to move a step for nigh on twenty minutes to the delight of twenty-four day-tripping Japanese students, who filmed and photographed me growing increasingly hot and bothered. Like the seven stages of grief, I ran the gamut of emotions. Shock and denial: *What, you won't go? No, no, you will.* Pain and guilt: *Why are you doing this to me?—Am I a terrible owner?* Anger and bargaining: *Giddy-up, you bugger. All right, how about for a carrot?* Depression: *That's that then. We're stuck here forever.* The upward turn: *Well, at least we're amusing someone.* Reconstruction and working through: *Is it you? Is it me? Is it us?*

And finally, acceptance and hope.

"Fine," I muttered, red-faced, sweating. "Let's go back home. We'll try again tomorrow."

Obedient as an aged Labrador, Major turned and walked back to the barn to a final rousing cheer from my audience.

"So many 'problem behaviors,'" Nahshon told me five years later in North Carolina, "come from animals who don't feel good about how they're treated. They are saying no and not being listened to."

Because our human ego, he said, does not like hearing no.

"The ego is afraid of feeling forgotten. And what are the ways we're forgotten? By not being seen or heard. When we hear no, we feel not heard. So we get bigger; we get louder. All because we're afraid of being forgotten. And there's no peace there."

He's right. That day, deathly embarrassed, my ego wanted compliance.

Did I care about the reason for Major's no?

Did I want to listen?

Not really.

I just wanted him to bloody well *move.*

Shawna Karrasch is a bright, bubbly veteran animal trainer whose early career training captive dolphins, orcas, and sea lions ("They're my favorite," she says. "They have lots of insecurities.") taught her to "embrace the no."

"You can never," she tells me over Zoom, "force a marine mammal to say yes. They get all their food and social interactions every day, regardless of what they do or don't do for you. If they don't want to participate, they'll just swim away."

Generally, she says, you get yeses by being fun and by building your relationship through training. But the *nos* are where your human learning happens.

"If an animal says no, instead of taking it personally, ask yourself: *Why*? If you're training a new behavior and your learner's just not getting it, you might need to back it up, make it clearer, help them *find* the right answer to feel successful."

But there are, she says, manifold other reasons for a no.

"It might be 'No, I can't,' through pain or lack of ability. It could be 'No, I'm afraid,' or 'No, I'm not ready yet, but maybe later,' 'No, this isn't fun or interesting,' or 'No, my needs aren't being met.'"

This reminds me of something Ronald the cowmunicator told me.

"Often," he said, "I'm contacted by farmers who'll say 'My cows don't want to go out. They don't like it outdoors.' But if I look in detail, I see very silly things. Last week, I went to a farm with no shade outside and with the drinking water all indoors. Or there's a shadow in the barn doorway, which for a cow is a black hole that she'll refuse to walk over. Put up an extra light, and presto—problem solved."

But there's one more possible reason.

"A no," Shawna says, "may come from a place of trauma," something she herself understands well. "I was sexually abused from age three to sixteen and grew up with severe dysfunction. Whenever a problem arose in my life, I didn't want to deal with it. Instead I ran."

Much later in life, she says, finally dealing with those 'raw hard feelings' helped her understand the nos of others.

"I've worked with traumatized animals who want to kill people. And it's not personal. They're not doing things *at* us, *against* us, *in defiance* of us. Animals' prefrontal cortexes aren't developed enough to include malice. Instead, I think: 'You are simply communicating that something's amiss with you. How can I help you move forward from it in a kind, clear way?' When you look through a trauma-informed lens, it's easier. But it takes clearing your own self up first."

Sometimes, though, Shawna says, it's none of the above. "Sometimes your animal might just be having an off day. We all get up some days and don't want to do the things we love. All right, no big deal. Either go away and come back later, or pick something easier to work on, and see if you can't get them back in the game."

"And if they *still* say no?" I ask.

"That's OK. Lose the leadership myth. The disrespect myth. If they're walking all over you, it's because they don't know not to. Ask yourself: What motivates the animal that I'm not doing? How am *I* not making it worthwhile? Flip that over, and you'll succeed."

She has seen, she says, hyenas, meercats, hippos, rhinos, sharks, giraffes, horses, and lions successfully trained by doing so.

"Is there," I ask Shawna, "any animal with a no so big that they're beyond hope of a yes?"

"No," she says firmly. "But I think there are humans who aren't yet patient or skillful enough to help them. The limitation is in the people, not the animals. Sometimes it's about unpacking ourselves, examining our own insufficiencies, and letting go of this agenda we've got riding underneath."

Or, I think, not riding at all, as two dozen merry exchange students once observed.

The English village of Newton St. Loe is the achingly bucolic sort of place it's hard to believe exists outside a Richard Curtis movie. Buttery stone cottages line flower-filled lanes, backed by woodland and meadows of grazing sheep. A farm shop dispenses ice cream to happy children and cream teas to their chatting parents. I park in its farmyard, glad for a break from my mother's attic. She's moving north to the country, so I'm on a trip home to sift through my old stuff. Beside the car, a dozen pink piglets cavort around a wallow. Bumblebees animate a bank of English lavender.

This place is old. It appeared in the Domesday Book of 1086. A Roman villa was discovered nearby, along with the remains of mammoths and prehistoric horses. And today, on this unusually scorching Wednesday, I'm here to visit a human-animal entanglement that itself dates back some eight thousand years.

Through a gateway beside the bumblebees, I enter the Bird of Prey Project, a conservation charity that advocates for the welfare of both wild and captive birds of prey. They do this with an unusual approach to the ancient art of falconry: by allowing, even *encouraging*, their "team" of twenty owls, falcons, and raptors to say no.

I arrive in time for the 2:00 p.m. flying display, led by center manager Naomi Johns and her deputy, Alice Davidson. An audience assembles on benches in a clearing. Charlie, a fourteen-year-old Harris's hawk, swoops in, returning from his free-flying "hawk walk," accompanying a group of human visitors.

The display's first participant, a lugger falcon named Indali, opts out of flying in such temperatures, preferring the cool of her day aviary. Instead, Milo, a tiny merlin, consensually dons his "backpack," a GPS-equipped harness that sits almost weightless beneath his feathers, and puts on an astonishing aerobatic display, so fleet he can catch dragonflies on the wing.

Next comes Finn, a long-eared owl whose wild relatives are threatened in Britain. If his ears lie flat, Alice explains, all's well. If they are up in tufts, he's either on the alert or sleeping: Like human gestures, she tells the audience, an animal's body language can mean wildly different things.

Curtailed by the heat, the display ends early. The crowd applauds. Naomi feeds Finn a treat and takes him home. She offers me a handshake, immediately retracts it, and wipes one viscera-smeared palm on her trouser leg. "Sorry." She grins. "You probably don't want to shake that."

I don't know much about raptors, though one of my favorite books is J. A. Baker's 1967 *The Peregrine*. "You cannot know what freedom means," he writes, "till you have seen a peregrine loosed into the warm spring sky to roam at will through all the far provinces of light." At home, I love watching red-tailed hawks and turkey vultures patrol our meadow. As a child, I was fascinated by the stuffed snowy owl who just gave me a fright in Mum's attic, which my great-grandfather shot "by accident." During the first desperate months of COVID 19, a young barred owl often accompanied Gal and me on our nightly woodland "sanity walks," once dangling a dead squirrel like a comfort blanket.

But I do know that captive working falcons traditionally wear hoods and jesses and leather creance leashes, collectively called "furniture," and that falconers have historically given them little choice in the training process, lest their inherent wildness causes them to fly off and never come back.

Hence, though many captive wild species these days learn through choice-based training, like Freddie and Freckles, a pair of black red-tailed cockatoos I met yesterday at the London Zoo, performing trained, free-flight behaviors for a crowd, birds of prey are considered somehow wilder, in need of "breaking," like wild mustangs, to ensure their cooperation.

"Developed in Europe over the last two thousand years," Naomi tells me, "modern falconry has a really long-standing way of doing things. But here, over the last eighteen months, we've completely changed our approach to using consent-based, high-trust training with our whole avian team. The idea and main motivator behind it is choice. Every action we ask of our birds is voluntary: everything from stepping on a weighing scale to going out to lead a display."

Take Indali. "Today, I opened her aviary hatch, and she chose to back inside, which means 'I don't want to perform, thank you.' We could coerce her to come and fly. But instead we respect her decision, and that's how we motivate her with choice."

Some people, Naomi says, maintain it's impossible to train a raptor through free choice. "I've had plenty of falcon keepers tell me I'm delusional. But when you put the choice into the birds' hands, so to speak, 99.9 percent of the time, they will say yes."

"What," I ask Naomi, "prompted this controversial change of approach?"

Self-taught from a young age, she says, she read hundreds of books on falconry before being mentored the traditional way.

"In its first few days of training, the bird learns, while tethered and under a certain amount of stress, to tolerate people until it feels comfortable to feed in the presence of that person and eventually flies toward them on a safety line, working toward one day flying free."

But Naomi had, she says, "some rumblings of an idea" after talking with a friend who trains pelicans, cranes, and magpie geese to perform voluntary cooperative care procedures, like walking through a medicated footbath or presenting a body part for medical inspection.

"It blew my mind that this was possible with birds. That night, I stayed up, brain whirring, and next morning, I went to Milo and cut off all his furniture. Within a week, I'd trained him to voluntarily raise his foot for exams, to accept a backpack for his GPS transmitter, and to put his beak into my fingers. During that week, my whole concept of these birds' intelligence, their theory of mind, their understanding of consequences, was shattered."

The pinnacle of the project's achievement so far is the "tame hack": the birds' daily unsupervised wild flying time. Casper the European kestrel, for example, goes out alone at 10:00 a.m. each morning, then returns at eleven for the day's first show. "In that time," says Naomi, "he flies six to twelve miles at over eight hundred feet in altitude." High-trust training, she tells me, is all about reading and understanding the minutiae of behavior. "Because if the bird is out flying free, there's absolutely nothing you can do if he decides 'That's it. Goodbye. I'm done here.' The relationship has to be strong and built on trust. And trust starts with respecting their decision to say no." Naomi knocks on wood. "We have never lost a bird yet."

Offering the birds choice, she tells me, has improved their daily lives in manifold ways.

"Some were tiny, marginal changes that you wouldn't notice unless you really knew them well. They started bathing more frequently. They were more relaxed between flying times. They sunbathed more. Casper started using a brand-new vocalization. It's no longer just a trade-off, exchanging food for doing a job. Choice has become the main motivator for all of us: knowing and doing what the bird wants."

Back to Indali, for example. "People call lugger falcons temperamental and said 'good luck with *that*' when I took this approach to training her. But now, working restraint free, I understand what those temper tantrums really are. Luggers are just extremely sensitive. Indali doesn't like it, for instance, if her routine's changed even slightly."

I accompany Naomi as she performs some cooperative care. Gently, Milo places one foot atop her finger so she can apply a dab of ointment. "This sort of thing really amazes a falconer, because it's the opposite of what we've been taught—that good training is a bird who simply tolerates things."

We visit Wee Man, a partially sighted peregrine who flies his solo tame hacks accompanied by a drone, allowing him freedom while ensuring he find home safely. Arwen, the enormous rescued European eagle owl, who was kept for three years in someone's living room and whom Alice is now teaching to express a choice. Florence, a tiny beetle-browed burrowing owl, whose long legs make it look as if she's pulled her socks right up.

"The project," Naomi says, "is still a work in progress for all its participants. We don't have all the answers. Some days I feel I don't know anything. But we're striving to better ourselves constantly. We're learning every day."

A classic case, I tell her, of "beginner's mind."

Naomi nods. "We're changing the definition of perfect welfare from a bird who looks visibly content to one we know is content. That's the true power of no."

Back home, Dolly has a "no" problem.

It's partly nature, partly nurture, with a sprinkle perhaps of early brain wiring. Her parents and her parents' parents were livestock guardians. Dolly considers her people her livestock. As a puppy, she bonded with us on the car ride home and has never needed further humans in her life. And though Turid told me that COVID-19 puppies are the "luckiest on the planet" ("People usually stress new dogs out by launching them too quickly into the outside world," she said. So when it comes to dogs, I say: Let's have COVID forever.), it meant we couldn't expose her to other people in our home, as we had our French mastiff, Agador, who endured raucous children's playdates and raucouser adult dinner parties.

Dolly tolerates people while out and about, though even then, like a cat, she's friendliest to those who display no interest in her. She will even go without bother into other people's homes. But park in our driveway or, horror of horrors, *come into the house*, and Dolly becomes the poster child for stranger danger. "No!" she barks. "No! No! No! No! No!"

Classic positive reinforcement, the sort of R+ treat training that worked wonders on Andvari and on zoo animals across the globe, doesn't function in those moments. She's too "over threshold" to care about my cheesy puffs. But summer's coming, and the kids are bringing friends to the farm. Were she a Yorkshire terrier, I might just tuck her under one arm and ignore it. But she's huge. And intimidating. And will scare off all the eleven-year-olds.

What else can I do?

I search online, through reams of behaviorists offering shock collars and prong collars and incredible ten-minute fixes, and come across an astonishing video. A camera moves slowly down the wire-and-concrete aisle of a municipal dog shelter. It's a familiar sight, except that every dog is sitting down or lying calmly in its bed. No whining. No barking. No frenzied flinging of furry bodies against the kennel bars.

I send an email and arrange a call with the people responsible for this alchemy: behaviorists Maasa Nishimuta and Sean Wills, who use a modality called, delightfully, constructional affection training—or CAT, for dogs.

Sean and Maasa are based in central Florida, "just south," says Sean, "of where the rockets launch." We meet just thirty minutes after they've brought home their new rescue dog, Remmy, a two-and-a-half-year-old boxer-pit bull mix with an adorable underbite. He's settling in, sniffing about their office. As we chat, his tail appears on-screen, a black-and-white cornstalk waving in the wind.

Sean is laid-back and funny, Maasa a ball of infectious energy. Both animal lovers from childhood, Maasa grew up in Japan, where, she says, she spent her childhood petting street cats, "or 'committee cats,' as we called them. People from this house would feed them breakfast; people from that house would feed them dinner." Together they now run the nonprofit Constructional Approach to Animal Welfare and Training, which offers training and behavioral support.

How, I ask, should I go about turning Dolly's no into a yes?

Take the constructional approach, they tell me. Pioneered by their former professor Dr. Jesús Rosales-Ruiz and his then graduate student Kellie Snyder, this entails simply *constructing a new behavior* (Dolly being pleased to see strangers) instead of trying to *deconstruct an existing one* (Dolly being highly displeased about seeing them). This works, they explain, both for domestic and captive wild species, where fear or aggression makes positive-reinforcement cheesy puffs untenable. "The cornerstone," says Sean, "is to identify what your learner wants in that moment, and teach them how to get it."

In illustration, they tell me about an "unadoptable" dog named Skip who'd been resident at their local shelter for over two years.

"He was very dog-reactive," Sean begins, "so much so that when staff gave him trazodone, an anti-aggression medication, his aggression would actually get worse. So much so that if a human were holding his leash, he'd redirect and start attacking that human."

"So we needed," Maasa takes up the story, "to teach him how to interact with dogs. Whenever we'd walk up to him with a dog, as soon as he *looked* at us, well before he started barking or lunging, we'd walk the dog away and give him space. We repeated that over and over. What we were teaching him is that you don't have to freak out, because if you just *look* at that dog, she'll leave you alone. That's the first step to the constructional approach."

Skip, Sean explains, was learning that with his own actions, he could have control over his environment. "And once he had control over it, it took away the scariness triggering his reactivity."

Eventually Skip reached a "switchover" moment, where his fear became curiosity. (This phenomenon, which I call the magical switcheroo, is how I taught Andvari not to fear the rainbow umbrella: show him a weird object and retreat before he does. Invariably, when I present it again, he's curious. Repeat, and like Alice, he's curiouser and curiouser.)

"Over the course of a couple of months using that method," says Maasa, "Skip let the dog come closer and closer, became friendly with that dog, and then with other dogs. In another few months, he became friendly with people—and he was adopted."

"We still get texts from his adoptive family," says Sean. "He goes to the beach. He sleeps in the bed. He's got it made."

But the constructional method, Maasa tells me, is not just for household species. "It's *generalizable*."

Their colleague Lisa Clifton-Bumpus, an animal trainer who also learned from Jesús, has used it to teach rhinos to accept injections and a giraffe to perform its own chiropractic adjustments.

"Her work is nothing short of jaw-dropping," Sean says. "She has revolutionized captive giraffe care."

Likewise, one of their own "star pupils," Kyle Hetzel, who taught an African bull, a kookaburra, and a sea lion to voluntarily undergo veterinary procedures. At first, each answered a resounding no. With time, the magical switcheroo worked its wonders. Kyle's results, Maasa says, were "spectacular."

Sean and Maasa themselves, meanwhile, have developed a specifically canine variation on this theme: the delightful constructional affection training for ebullient dogs like Puff who can't help but jump up for attention.

"Our goal is to get dogs in shelters adopted," Sean explains. "Behaviorist Dr. Alexandra Protopopova looked at five hundred adoption interactions in northern Florida and found that only one behavior led to a statistically significant increase in successful adoption."

It wasn't following commands, says Sean. It wasn't playing fetch, says Maasa. The *only* thing a dog could do to get themselves adopted, they tell me, was to approach a potential adopter and lie down beside them.

"If a dog does this at a meet-and-greet," Sean says, "they're *fourteen times more likely* to get adopted than a dog that does any other behavior at all. Holy crap. That blew our minds."

"Wow," I contribute, mind blown also.

At first, he continues, they tried using treat-based positive reinforcement to train shelter dogs to lie down calmly.

"But shelters are loud and busy. And getting the dogs' attention was tough because they weren't deprived of food or treats. What they *were* deprived of was interaction. That's when it hit us: Maybe there's a different kind of training we can give them."

The result, constructional affection, was developed by Sean, his friend Chase Owens, and Dr. Jesús and is responsible for that incredible scene of shelter contentment I first spied online. "We got to implement that program to revamp the entire facility," Sean tells me.

The key to calming an overexcited dog, Maasa and Sean explain, is first to understand that for that individual dog, *affection* is the best reinforcer.

"Any human who finds themselves being mouthed and jumped on and their clothes being pawed should really be thinking: *They love me so much*," says Sean. "It's the biggest compliment a living thing can give you."

But shaping all that love into something less chaotic, they say, takes just four simple steps. First, if your dog approaches you for attention with all four feet on the ground, pet her with one hand. Second, if she jumps up, stop petting. Third, when all four feet return to the ground, continue petting her with one hand.

"And fourth," Sean instructs, "if she sits or lies down, pet her exuberantly with both hands. Really lavish on the praise. We say we turn ourselves into affection vending machines for the dogs; they can come up and select what they'd like. It's still positive reinforcement, but it's all about how you deliver that reinforcer."

It seems to be working for new arrival Remmy, who, after receiving great dollops of constructional affection, has now settled down to nap between them.

This approach, like the first, extends far beyond dogs. One trainer in Japan, Maasa says, decided to use it with her tortoises.

"Now she has traffic jams when her tortoises all race to her from across the room. They crawl into her lap, and she just sits there, petting the whole pile of them."

Sean found something similar while interacting at a zoo with another—giant—tortoise. "When I started petting his shell, he'd wiggle and push into the pressure. So I asked the zoo, 'Have you noticed he likes to be petted?' No, they told me, but a tortoise's spinal cord *is* attached to its shell, so he can definitely feel it. He grew to love it so much that he'd start ramming his head against my leg if I stopped petting him to talk to people."

We might assume, Sean concludes, that for solitary creatures such as tortoises or reptiles, interaction through touch isn't as important as to social species. "But the eye-opening thing is it's exactly the same."

Touch was certainly the key, in Andvari's panic-stricken early days, to his little fluffy heart. But first I had to teach him belief in the power of saying no.

It was clear, from the first, that he'd never been permitted to say it. But "yes only means yes," I remembered Warwick saying, "if no is a genuine possibility."

So I taught him that if he said no, I would respect it. At first, I could tell he doubted me; he waited, tense, for some consequence or other. But slowly he came to believe I'd keep my word.

When you first give an animal the option to say no, you tend to find the nos come thick and fast.

For a while, our conversations went something like this:

"Do you want to go out for a ride?"

No.

"Well, how about a walk? Side by side?"

No.

"Want to be groomed?"

No.

"Scratched?"

No.

"Well, would you at least like to be touched?"

No!

Hard as it was for human, ego-burdened me, I honored them.

Then eventually came the magical switcheroo.

"Would you like to be touched?"

Yes.

"Groomed?"

OK then.

"Scratched?"

Right here, please.

Scratching, Andvari's constructional affection, sealed the deal. It was April by then, when Icelandics shed their inch-thick Christmas sweater. With this comes profound itchiness, as if they were suddenly wearing that sweater with nothing underneath.

So with his permission, I'd dig in my nails to his shoulder, his chest, beneath his chin. The reward: a wobble-lipped, clacky-teethed jelly of a pony. *Down a bit*, he'd crane his neck and tilt his head. *Up a bit… Left… No, not that left. That's it—just there!* A chin wobble. An eyelid flutter. Then: *clack-clack-clack-clack.*

I'd pause.

Gently, he'd turn and nudge me.

More more more, he was saying.

Now I start experimenting with Dolly.

I pay the boys' teenaged friends to come over, park on the driveway, and approach.

Together we dance around the tender edges of Dolly's comfort zone: a zone—a sort of canine KerPlunk. We stay where she's calm: well below threshold. When she looks at me or away or sniffs something or yawns, the obliging teenaged visitor walks away. We repeat and repeat. We err on the side of caution. Never once do we accidentally push too far.

Within a dozen or so tries, they can each walk right up to receive their ten-dollar payment.

Then, to make the most of my money, I send them inside to see Puff.

Feet up: No petting, I instruct them. Feet on the ground: one hand. Sitting or lying down: all the delight you can throw at her.

From the window, I watch as she wiggles and twists, then plants her bulldog bottom in a valiant attempt at self-control. She jumps up, spins a circle, grabs a shoe, lies down, jumps up, and then sits again. Every grain of her being tries to do the right thing. She can. She can't. She can again.

The results, though expectedly unlinear (Skip, after all, took months) are, as Sean presaged, spectacular.

I vow to keep at it till the ten-dollar bills run dry, just as I did ten years ago with a chicken named after a bottle of gin.

One afternoon in 2015, I received a frenzied phone call from a friend. "There's a hen who needs a home by six o'clock," she gasped, "or she's going to be slaughtered. I can't take her, but can you?"

Roughly two hundred million chickens are killed, worldwide, by humans each day—the equivalent to five times the human population of California or three times that of Great Britain. Perhaps then this one anonymous life wasn't all that important. But I've always admired what Wordsworth, back at Tintern Abbey, called "little, nameless, unremembered, acts of kindness." Hence off I traipsed, stopping at our local liquor store to beg a box in which to transport the little, nameless rescue.

When I arrived, I saw, cowering in the farthest, filthiest reaches of an otherwise empty coop, a sad and sorry specimen. Her owner had, she said, dispersed her flock, but nobody wanted this one last hen: a tiny, bedraggled, off-white Silkie bantam. Terrorized by an overenthusiastic rooster and clearly at rock bottom of the flock's pecking order, it seemed she'd never had a positive interaction with another chicken, let alone a human.

"Take it if you want," shrugged the owner. "Or I'll kill it later."

I eyed the owner. I eyed the chicken. And with difficulty, heaving from the cecal stench, I caught and bundled the petrified latter into my Bombay Sapphire box.

"I'll call you Sapphire," I told the box as I drove home.

Very gradually, over the next few weeks, I trained my new old chicken to be happier. First, I kept her in a dog crate in a dark, quiet corner under the stairs. Every few hours, I arrived to feed and water her; at her first inevitable sign of fear, I would retreat. Slowly her tolerance for me, then for the rest of my family and other animals, increased. She stopped hitting the roof, literally, at every noise or movement. I moved her to her very own coop outdoors, where she could scratch and peck and eat unmolested. And eventually, months later, came the magical switcheroo. Sapphire started seeking out attention. She would sit cuddled on my lap, beg for treats at my feet, announce her pleasure while taking a dust bath in the sunshine. I found her a friend, another rescued Silkie we named Jubilee. They moved in together.

"Connection before concepts," writes psychotherapist Sarah Schlote, who works in horse-human trauma recovery. "Attunement before asking. Partnership before procedures." I wasn't thinking much then about concepts or procedures. I didn't know what CAT was, or the four quadrants of operant conditioning. I don't think I even really knew I was training Sapphire. I was just trying, very gradually, to change the nos to yeses so she could live in peace.

Which she did, with us and Jubilee, for the rest of her life.

At 7:00 a.m. on May 18, I am blearily making coffee when something zips past the kitchen window. I rub my eyes and look again.

The hummingbirds are back.

I rush for the feeder, boil sugar water, let it cool, and set it out. Within moments, a little male is there, hovering, drinking, zooming a neat figure eight, and returning for more.

Mervyn? Could it be that this tiny miracle of natural engineering has successfully navigated oceans and predators and starvation on the wing to make it back to his very own farm and his very own feeder?

He hurtles off and alights on the topmost raspberry cane.

His repose, however, is brief. Within moments, another male has appeared, and Mervyn is chasing him away. Even after a harrowing migration, a hummingbird's work is never done.

Siegfried, I am certain, would understand Mervyn's toil. He spends his days in the wood, fending off interlopers intent on poaching a wife. Then as afternoon lengthens, he struts to the very center of the meadow. What triggers this daily odyssey? The arrival, it seems, of a pair of Canada geese who glide in, honking noisy copilot calls, late each afternoon to graze in the meadow's marshiest dip. Zeyah names them Howard and Penny. Like pigeons, Canada geese mate for life. The pair peck and preen and fall asleep, necks entwined like pottery swans.

Something about them affronts yet intrigues Siegfried. He bursts out of the wood, flapping and crowing, but stays a good twenty yards away. There he stops and stares. And stares. And stares.

Howard and Penny ignore him.

But still he stands there, unable to look away.

I am watching him watching them when Gal comes to find me.

"Where did you hide the pigs?" he says.

"Who?" I look up.

"Constance and Agnes. Where did you put them?"

"*What?*"

I scramble up and fly to the garden, where, sure enough, an excavated portion of fencing tells of Escape from Porkatraz.

I race to the kitchen door.

"Quick! Everyone!" I yell in. "Find the pigs!"

Indy runs outside. Zeyah doesn't even put on shoes. We scatter.

"Constance, Agnes! Here, piggy-piggy-piggies!"

I wait. Listen. Nothing. Not an oink, not a snuffle.

What, I catastrophize, if we can't find them? I imagine them on our country road. The daily milk wagon on its way back to the depot… The squeal of pigs… The squeal of brakes…

"Boys?" I holler. "Anything?"

From the far-distant edge of the orchard comes a faint "Found them!"

Think…think… What will garner a yes answer? What do they want? I rush to the kitchen cupboard, grab a metal mixing bowl, throw in handfuls of grain.

I sprint with it across the orchard.

Up ahead, I see one husband, one eleven-year-old, and two swineagers receding rapidly into the distance. "Overtake! Overtake!" I cry.

They intercept the fugitives just before our neighbor's farm shop. Breathless, I catch up. I glance down at my mixing bowl. Will it be enough? We haven't commenced leash training yet, and it's a good quarter mile, avoiding the road, to get home. They've grown too heavy for us to pick them up. We have no crates big enough to contain them.

Zeyah arrives and tries to slip a dog harness onto Agnes. She screams her indignation, then spins like a short, fat dervish until, with one definitive moonwalk, she backs right out of it.

A yes answer, I realize, is all we've got.

I rattle my bowl. "Ladies?" I venture. "Want to come with me?"

Agnes wags her tail. Constance snuffles my sneaker. Merrily, they follow the porcine Pied Piper.

"We're trying to get the pigs home. We're trying to get the pigs home. Are you coming with me?" I chant as, shaking my bowl in time to the beat, I lead them back through the neighbor's farmyard and down the lane to the river.

All along the rail trail, trit-trot, trit-trot.

Up through the meadow, swishy-swashy, swishy-swashy.

Straight past the horses, snort-snort, snort-snort.

Back through the orchard, nearly there, nearly there…

To great shrieks of protest, a new one for the pigtionary, I shove first bowl, then two red bottoms unceremoniously under the garden fence, like toothpaste back into the tube.

Panting, we all collapse on the grass.

Behind the fence, the pigs answer "yes" to their afternoon snack.

I go inside for a tape measure. Time to order those harnesses.

It wasn't long after the day Major refused to go forward that he began walking backward instead.

"Come on," I'd say, urging him along.

No, he'd reply, and then back the truck up.

He no longer wanted to go on our long, languid trail rides. He got grumpy and nippy. He went lame first on one foot and then on another.

Years of trying to get to the bottom of it ensued.

We tried X-rays and ultrasounds. Injections. Medicines. Supplements. We tried corrective shoeing, with plastic pads and pour-in gels. We tried exercise. Rest. No shoes at all. He saw a chiropractor. I trained in massage and bodywork. Things helped for a little while, and then they didn't.

So I listened to the no, stopped riding, and gave him the choice. On good days, "Would you like to?" I'd ask. If he agreed, we'd saddle up and go for a gentle wander. On bad, he'd limp about, resting one foot, then the other. "That's OK," I'd say. "Let's just spend time and try to make you comfortable."

No is not always uncategorical. No is not necessarily forever. But neither is yes. And you can only know an animal's yes is a true one if no is an answer you are willing to hear and accept. In a shelter situation, the yes animals almost always get adopted first. But there's always hope for the nos: for the ones who say it quietly and the ones who say it loud. Because when you help an animal go from no to yes, your relationship supercharges.

In part 32 of Whitman's "Song of Myself," he writes about animals:

> They do not sweat and whine about their
> condition,
> They do not lie awake in the dark and weep for
> their sins,
> They do not make me sick discussing their duty
> to God,
> Not one is dissatisfied, not one is demented with
> the mania of owning things,
> Not one kneels to another, nor to his kind that
> lived thousands of years ago,
> Not one is respectable or unhappy over the
> whole earth.

Every being, I believe, wants to be heard and to be happy. Instead of an affront or rejection, consider no a yes *to something else*: a suggestion or a change of subject. Then ask yourself, what's *that* yes all about? What's in it for them? What can I change about my question or the way I ask it to make it something they want too? If you're training, you might need to appraise your methods, whether they help you listen or talk over the top. Or you might need a long, hard look, uncomfortable as that can be, in the looking glass. Or perhaps you need to seek out the fun, the play, the biology of joy.

We are all born into that biology of joy, before life leads us away from it, whether we're an abandoned dog, an anxious pony, a ball-pushing bumblebee, or a neglected chicken. Understanding another being's no can be the key to them finding their way back.

It's been three years now since my last "Would you like to?" But today the sun is shining, the air's warm, and Major looks sound and shiny. I fetch bridle, reins, and his black saddle, festooned with spiderwebs from its winter in the workshop.

Major leaves his hay and strolls over. He sniffs the array of equipment.

"You want to?" I offer his bridle.

He reaches his big red head forward and gently picks up the bit between his teeth.

"What about this one?"

He cocks one foot as I place a thick felt saddle pad on his back.

"And this?"

I swing his saddle up on top.

Interested, Major turns his head to inspect it.

I buckle the girth, first loosely, then tighter. Clip on the reins. Give him a couple of mouthfuls of hay.

"What shall we do now?" I ask.

Major turns again and glances meaningfully at the stirrup.

I look down at my wellies, at least three inches too wide to fit inside. "Don't tell anyone, all right?"

I slip them off, place one pink-socked foot in the stirrup, and swing myself gently up onto his back.

He blinks and chews his hay.

Up here, the world feels glorious. I hear Sven chitter from the wood. The pigeons loop over in loose, bunting swags. I remember our rides of the past. Those gallops through the forest, so fast it felt like flying, when the only words I could summon came from Shakespeare's *Henry V*: "When I bestride him, I soar." I let that longing pass by like a soap bubble, try to concentrate on this moment, and feel an intense rush of joy.

He offered this, I think. He could have said "Eff off." Could have changed his mind. Changed the subject. But he didn't.

Now here we stand in the sunshine together, me in the saddle, he beneath it.

"That's enough for today." I grin and hop off, plunging socks into sticky black mud.

Major sighs. He closes his eyes and falls asleep.

I stand with his chin in my hands until he wakes up.

PART III

The Feelers

The Art of Listening to Animals

in · tu · i · tion /ˌintŏŏ'iSH(ə)n, ˌintəwiSH(ə)n/

> *noun*
>
> a thing that one knows or considers likely from instinctive feeling rather than conscious reasoning.

10

Back on Track

Listening Through Attunement to the Natural World

> "Several of Nature's People
> I know, and they know me;
> I feel for them a transport
> Of cordiality."
>
> —EMILY DICKINSON, "A narrow Fellow in the Grass"

JUNE

Summer comes in with a wallop. It's hot and sweaty, and open doors mean fewer barriers between the human world and the rest of it. I extract an Eastern red-backed salamander from the recycling bin and reroute an inch-long seed bug from his morning constitutional up our bedroom curtains.

Goldfinches descend on the orchard like glitter. Mervyn darts in and out of the raspberry patch, perhaps feeding fledglings the size of fat blueberries. The dawn chorus is in full voice, though Siegfried has gone silent, I suppose for tactical reasons. Pheasants nest on ground level, a risky business. Incubating a dozen camo-colored eggs for about four weeks, it pays to contain your shrieks and shouts for the duration.

At least I hope that's why.

On June 1, Indy and I find a clump of gray-brown striped feathers at the edge of the wood.

"Do you…do you think…" His face falls.

"I don't know. I hope not."

We comb the grass for evidence and happily find none.

It reminds me of the trauma of keeping chickens. How an unleashed dog pushed its way into our garden and killed Marigold, our little orange hen, in front of Indy; how he cried for a week and gave her a burial festooned in flowers.

Save for Siegfried, though, nature has gone into overdrive. The world teems with life, with voices—so many that it's hard to tell what's what, who's who, what anyone's saying. I return, one more time, to Dr. Jenifer Zeligs. Scientists, she said, learn with the goal of knowing. Trainers learn with the goal of doing. "But intuitives, like trackers and animal communicators," she told me, "are interested in the value and interconnection of all beings, in feeling our common 'beingness.'"

Scientists, I consider, have helped me find out what there is to hear.

Trainers have shown me better ways to listen.

But what about the intuitives? Can they, I want to know, teach me to *feel* more too? Is intuition the missing piece that would allow me to understand all these interconnected individuals within this one gorgeous groundswell of life?

I spot a poster advertising a course at a beach not far away, held by master holistic tracker Arnaud Gagné, in one week's time. "Holistic," the old-faithful dictionary says. "Characterized by the belief that the parts of something are interconnected and can be explained only by reference to the whole." This seems a good place to start. I email, and he invites me along.

In the meantime, I work at untangling the multitudinous sounds around me. Outdoors at dawn, as the sun rises hot and heavy and starlings snap, crackle, and pop in the oak, I take wildlife tracker Josh

Lane's online "Bird Language" course. Birds, Josh says, have five voices or basic vocalizations. Aside from song—that stirring combination of "This is me," "This is mine," "Hello ladies," and "Ain't life grand"— there are companion calls ("I'm here") and territorial aggression calls (wild ravens, I find out, have at least seventy-nine unique kinds of these). There are baby begging calls, in which our newborn starlings specialized, and alarm calls: "Watch out!"'s that descend into what Josh calls a "pillow of silence" as birds wait for the threat to pass on.

As I perform my morning chores, I try discerning the baseline calls of relaxed birds, the urgent alarm calls, the sudden hush when a hawk sweeps over. I find that often *my* arrival is the cause for alarm (Josh refers to this as the "observer effect": the phenomenon in physics whereby the observer disturbs the observed) and feel joy, a sense of accomplishment even, when the baseline resumes.

I suppose I'd always thought each bird was just its own little one-way radio, broadcasting out into the world. Now I start to hear cause and effect, realize we're piggy-in-the-middling of conversations all the time. Pheasants, I read, have specialized calls for distress, mating, alarm, and incubation. Females use one call to warn their chicks of danger and another to gather them back in from hiding. I listen for them, but hear none, and hope Siegfried, Hildegard, and Fräulein aren't gone—just hidden too.

Josh's next exercise is to practice "clock listening." Sit quietly, he instructs, and identify what you hear, first in front of you at twelve o'clock, then to your immediate right at three, directly behind you at six o'clock, and to your left at nine. On my first round, I'm pathetic. At twelve, I clock a song sparrow and some *chickadee-dee-dee-dee-dees*. At three, the military tattoo of a pileated woodpecker. At six, directly behind me, a crow cawing. At nine, not much; the house seems to block out most sound.

I try again. Listen harder. Twelve, a twittering pair of bald eagles.

Three, a tractor: No, wait—it's the lawn-mower hum of a single engine plane. Six, far-distant traffic and the soft snorts of Frieda and Mary, the two Clydesdales who live across the road. At about eleven, the roar of a jet engine from forty thousand feet.

I do this twice daily. After a week of it, I can hear a beetle's *click-click-click* across the picnic table at 4:30 p.m. and almost feel my ears swiveling.

June 8 arrives, a sublime summer's morning, and I'm standing with three people on a wooden boardwalk that leads over a low sand dune. There's holistic tracker Arnaud, wearing gladiator sandals and his hair in a long, thin braid. Seventy-something Ann, who, as a child, spent summers in a pioneer cabin in the Colorado mountains. And Jesse, a middle-aged surfer with an interest in shamanism. We are all staring at a small, squashed pile of shit.

The sea air holds a maiden-aunt whiff of wild rose. A gull wheels and cries over the retreating Atlantic. A family settles at the base of the dune with beach towels and a plastic cooler.

With a twig, Arnaud pries the little pile apart. "What do we see?" he asks.

We all gaze.

"Excuse me. Sorry…" A young beachgoer squeezes past us, trying not to stare.

"We're just examining this poop," Arnaud greets her.

"Oh… Great." She flashes him a smile and hastens on.

Slowly, with probing, the answer emerges. Several small strands of hair. A yellow sliver of bone. The opalescent skin of an undigested rosehip.

"Tracking," says Arnaud as he pokes about, "brings conversation to the next level. It leads us on a journey of discovery into one moment in an animal's life."

This moment, we deduce, involved the early-morning excretion by a coyote (for coyote shit it is, evidenced—Arnaud gives us "Shit 101"—by its twisted, tubular construction) of the remains of last night's dinner: a mouse or vole or shrew, a bite of bony snowshoe hare, rosehips rich in vitamin C.

"And why," he asks, "is this poop in the middle of the trail?"

"Communication, maybe?" suggests Jesse. "'This trail is mine'?"

"Could be," Arnaud nods. He quotes Tom Brown Jr., the American tracker and survivalist under whom his own mentor, well-known tracker Jon Young, studied. "'The mystery reveals itself slowly, track by track.'"

We follow the coyote's probable path along the boardwalk, and—there!—find one paw print, then another, tracing a path down the dune to the beach.

We examine the prints closely. A tighter pad than that of a domestic dog and with four sharper nails. The sand around them is stippled with raindrop impressions, but the sand inside the footprints is not. We conjecture that these tracks were made sometime after 4:00 a.m.: "That's when the morning rain stopped," says Ann, who overnighted here in her VW campervan.

A crepuscular animal, on the move around dawn.

We spend the morning sometimes on hands and knees, sometimes on tiptoes, investigating and imagining the private life of this one coyote. Coyotes, Arnaud says, are edge dwellers. They stay out of the spotlight. I imagine this one, out here alone before the picnickers and the would-be trackers, trotting along on wet sand, at one with the morning. Was he here for a meal of quahog—hard clams—we wonder? We scour the tideline and find crunched shells. Or was this a shortcut back to his daytime territory? We mount a dune and look for possible paths and locations. Was he an outlier, I wonder, a brave earwig? Was he pursuing a scent?

I start feeling closer to this coyote I've never met. The beauty of tracking, Arnaud says, is that you learn to look without knowing. You trust your impressions, your intuition. You work from a quiet mind, moving your thoughts out of the way.

At midday, we station ourselves at the north end of the beach beside a strong-running stream that sweeps down from a marsh. "Before all else, the First Nations say, you must start with gratitude." Arnaud unwraps his sandwich. "Being grateful creates a softer way of being. I invite you each to express what you're grateful for."

Immediately my mind empties. This sort of group activity is already challenging my own edge-dwelling comfort zone. "We'll approach today," Arnaud prefaced earlier, "like children playing." But I sat out of musical chairs as a child and have religiously avoided all team-building exercises ever since.

Everyone looks at me. Ann nods encouragingly.

"I…um…"

Before us, a flock of dowitchers sweeps low over the stream in perfect synchrony. They turn, fall, rise, bank a sharp east, then repeat the choreography.

"I'm grateful"—I point—"for those…dowitchers?"

There's a pause. I hold my breath. Then everyone beams. They move along.

I look out over the beach and the forest behind us. These are the ancestral lands of the Mi'kmaq people. For thousands of years, I reflect, the Mi'kmaq tracked moose and wolves here; now there's just deer and coyote and people grateful for dowitchers.

Over lunch, Arnaud gives us tracking tips. We should each choose, he suggests, a "sit spot" within a one-hundred-footstep radius of our home that we can get to know intimately. "Go to that spot daily, and you'll become part of it. This is excellent for honing your powers of observation." Ann's daughter, a cartographer, she tells us calls this "ground truthing;" Arnaud calls it "dirt time." Edge habitats are good for this, he says: Food, and therefore life, exists at the margins. But most important, he says, is that it must be close. "Too far, and you won't go there enough. That's just how humans are."

We polish off our packed lunches. Time, Arnaud announces, for our "embodiment exercises." "We spent the morning thinking like animals," he says. "This afternoon, we're going to move like them, starting with coyote. Take off your shoes."

Oh *fuck*. Embodying the animal? *I am British*, I feel like telling him. We can think of no greater horror than being compelled to bay at the moon or flit like a butterfly. We don't even like taking our shoes off. Frantically, I pretend to jot down notes.

"Put down your pen," exhorts Arnaud. "Come let your body write the story."

I summon a queasy smile, write that down too, and reluctantly rejoin the group.

On all fours, Arnaud shows us how to move like a coyote: walking, trotting, running. Then, "Your turn," he says.

I try. I really do. But I'm about as far from coyote as can be. I can't tell which gait is which, which hand or foot to let fall in what order. I lumber along, fall over, feel a fool. I notice Arnaud snapping photos. It's so embarrassing that I decide not to think about what I'm doing. Yet the moment I let my thoughts go, I feel my body slip into it. I forget to be self-conscious and sense a cadence emerging. I have one fleeting, heart-lifting glimpse of how it might be to see this world through our coyote's eyes—low, straight, steady—and think of Ronald the cowmunicator advising me to look at things physically from the animal's perspective. For several seconds, I've moved my mind out of the way. I'm trusting my intuition. And then it's gone, and I'm just an idiot on all fours with sand in my socks once more.

Later, the group wanders deep into the marsh to track fox and heron prints in the thick, wet sand. But it's approaching five, and I need to get back for my literal dirt time. There is shit to be shoveled at home. We say our goodbyes. Arnaud manages to extract from me a final gratitude. "I'm grateful for our coyote," I say and mean it.

The group presses on. I turn to head back to the car park. But which way was it?

I follow the footprints.

George Bumann, artist, naturalist, and animal-language specialist, didn't start out your typical animal lover.

"I grew up," he tells me on Zoom the next morning, "a hunter and trapper and fisherman. Very honestly, I've killed more wild animals than most people have ever seen."

Nowadays, though, he sculpts and studies them instead.

"But I don't think there's ever been a time in my life," George reflects, "when I was as sensitive to the wind as when I was a hunter, because the deer and the elk and the foxes all are. You have to step into their world. Because of that interactive background, learning to call geese and ducks, competing in turkey-calling contests, I learned to become animal."

I'm not sure I've had similar luck with those methods. A while ago, at a flea market, I picked up a vintage instructional record titled *How to Call Wild Turkey*, which boasts instructors like Harvey Graybill, "1970 Pa. State Champion," and Roger Latham, "Mr. Turkey Himself." Very privately, I've been attempting to emulate these luminaries emulating turkeys. I sound like a Wookiee or Foghorn Leghorn. If the dogs hear me trying, they look startled, then appalled, then start barking wildly. Imitation might be the sincerest form of flattery, but like walking like a coyote, I suspect it won't help me "become animal" too. Is there anything else, I want to know, I can do instead?

We humans, George replies, all carry a feeling that we're not really part of a place. "This sort of 'us-them' separation: We're *in* it, but we're not *of* it." His antidote, like Arnaud's, is to "go out and sit somewhere with no phone, no agenda, and just be present."

But George goes one step further. *Duration*, he insists, is all-important.

"You'll notice you don't hear much at all for the first half hour or forty-five minutes, and then animals and birds will start doing their thing again. If we've never sat anywhere quietly for longer than that, we've never seen the natural world in anything other than a state of chaos and disruption that our own presence has created."

Schedule in at least forty-five minutes of quiet time, he advises, the way you would schedule a doctor's appointment. Then just sit and listen, long enough that you're no longer the disturbance.

"When people first do that, I see them go, 'Oh shit.' The stage curtain has finally been lifted."

For the last twenty years, George has been teaching this to visitors at Yellowstone National Park. "But," he says, "you can do this on top of a high-rise. You can do it on your front stoop in London or Manhattan. And you *should*. It's important to learn to love and see the place you live more fully through what its nonhuman beings can tell you."

Transformative moments, he adds, don't have to include "big-ticket" bears or bison.

"One day, I went outside to get away from my computer. I covered twenty feet in an hour or so, focusing on the sensory stimuli: the breeze, the sun, the sounds of the wind and the grasshoppers. A song sparrow landed inches from my toe. I watched it gnash a wild ryegrass seed in its beak. And something shifted that I couldn't put my finger on."

It felt, he describes, as if the world were not just around him, but moving *through* him.

"The sunlight isn't just hitting my cheek; it's passing through my face. The breeze is passing through my body. I felt like all my senses were on high and plugged in simultaneously."

On his way back "to the damn computer," he saw three black-billed magpies, *Pica hudsonia*.

"The closest you usually get to magpies is about thirty yards. But I kept taking steps, and they didn't move. I got to where, incredibly, I was almost standing over them. They were making impish, intimate little squeaks, and I felt this vastness within that moment."

Finally, he tells me, his "thinking mind" kicked in.

"I thought, 'this is crazy; no one's going to believe this.' And at that precise moment, they all stopped, looked right at me, and took flight. I couldn't get within thirty yards of them again. The spell had been broken. What *was* that?"

George's eyes gleam, teary from the retelling. "I think that's a glimmer of our ancestral condition. We were animal. We are animal. We still can be animal. And I think when we try, we are granted access into that family again. And it feels really good."

Even when you're short on time, such glimpses are golden.

"Walk to the mailbox. Go to the park for ten minutes. It's the hardest thing to do in our busy lives, I know, but it only takes a moment for your awareness to be rotated away from our human obsession with our own humdrum and drama."

You might see, he says, the deer with the injured leg or the raven who's been picked on by the others.

"And you see that your story is their story, and your piece becomes a bit smaller. It might even disappear. For me, that perspective switch is huge: to get out of ourselves, to be part of something bigger."

A bank of daylilies has burst open overnight. They blare to the sky like jazz trumpets. The chipmunks like them too. I watch one lily getting shorter and shorter, accompanied by the sound of munching from the foliage below. The air is syrupy with apple blossom and rambling rose. An old farmer chugs by on his ancient Massey Ferguson. With one hand, he waves a greeting; with the other, he holds a cell phone to his ear.

Indy is out spotting insects. A bicolored sweat bee, the British racing green of an old Mini Cooper. A white-spotted sawyer beetle, waggling his antennae wildly, who looks like a *Star Wars* bantha. A masked hunter assassin bug, fierce predator of lacewings and wood bugs.

While cutting a trail through the wood, I discover a deer skull, bleached white, with one antler and teeth still attached. The teeth aren't worn; it must have been a young one. I get a jolt of joy when I identify a bird call. I obsess over the possible speed and direction of a footprint. I trace alarm calls moving from wood to river—but never Siegfried's. The trail cam, though we check it constantly, doesn't catch him either.

On June 12, I connect with Louis Liebenberg to learn more about this addictive art of tracking. Born in Cape Town, Louis is a tracker, Harvard University associate professor of human evolutionary biology, and founder of Cybertracker, an environment-monitoring software that helps Indigenous master trackers worldwide gather scientific data.

He possesses an infectious, childlike glee. "I was drawn in early," he tells me, "by the mystique of the Kalahari Desert and the amazing tracking skills of the Kalahari San, who still hunted with bow and arrow." The San are hunter-gatherers, bushmen whose ancestral territories span South Africa, Namibia, Lesotho, Angola, Zambia, and Botswana.

He was also fascinated, he says, by physics and mathematics. "When I was a teenager, I became intrigued by the Wallace paradox, the one area where naturalist Alfred Wallace differed from Charles Darwin."

Wallace, Louis elaborates, believed that if we assume early "primitive" hunter-gatherers didn't "do" science like modern Europeans, then evolution's natural selection can't explain how the human brain evolved to be able to undertake it.

"Wallace thought the answer lay in divine intervention. But my solution to that paradox is to turn it on its head and say the art of tracking involves the *same* scientific reasoning that's required to do physics and mathematics. In other words, the hunter-gatherer tracker *is* a scientist, and the art of tracking may well be the origin of science itself."

At first, Louis's interest in tracking was purely personal. "In 1985, I dropped out of university to become a tracker. I went off alone in my old Land Cruiser into the Kalahari: no paved roads, no cell phones."

But there he met a San master tracker named !Nate, who both taught him practical tracking and explained his community's struggle to survive.

"In Botswana, cattle-farm fences had cut off the migration routes of wildebeest and caused a massive wildlife die-off, which resulted in hunter-gatherers not being able to sustain themselves."

Apartheid rule, moreover, had removed the San from their traditional lands for the creation of national parks with shoot-on-sight policies for "trespassers."

"Since then, many San had been living in shanty towns on their edges."

For traditional tracking skills to survive at all, Louis realized, they had to be developed into a paying profession. "The barrier, though, is that even today, some of the best Indigenous master trackers don't read or write. And that stimulated my creation of Cybertracker, scientific data-gathering software that doesn't require either. Its primary objective: job creation for Indigenous communities."

Like Western science, Louis tells me, Indigenous trackers employ both inductive and deductive reasoning.

"Say a group of San out hunting finds an indistinct track. Half might say honey badger, and half porcupine. They'll have lengthy debates based on their previous experience and on this really fractional spoor. They'll go up and down the trail, looking for more evidence with which to convince the others. That's evidence-based critical discussion and peer review at its most fundamental level."

Good tracking, Louis continues, also requires imagination. Trackers must project themselves into the place of the animal. And science, he says, is the same.

"A hypothesis isn't necessarily formed by examining objective facts: It's a question, an idea, a product of the human imagination, which we must then test, explore, consider, debate, and test again before we can prove or disprove it. Nature, to the tracker, is not an open book, but rather it requires him or her to read between the lines. Trackers can't read everything *in* the sand; they read *into* the sand. Just like scientists."

Master Indigenous trackers, he adds, are quick to admit their mistakes. "Ego won't get you the meat you're hunting. North Americans learning tracking,"—he smiles—"aren't always as good at this. There's always one smart-arse who refuses to admit he's wrong."

Tracking, he says, might be the oldest iteration of science, but that doesn't mean its place in the modern world is only for those seeking an escape to more primitive times. San trackers now monitor wildlife. They help intercept poachers before they do harm. They assist ecotourists hoping to spot the "Big Five" safari species: lion, leopard, rhino, buffalo, and elephant. They discover life in seemingly desolate places. They gather scientific data on wildlife populations, proliferation, migration routes, and behavior.

But there's also, Louis says, a more spiritual side to tracking.

"I've experienced this myself very fleetingly on a San persistence hunt, where hunters pursue the animal to the point of exhaustion. You interpret the tracks and signs. You visualize the animal. Then, while you're running it down, the combination of rhythmic running, intense concentration, and eventual exhaustion induces in you a trance state without drugs."

He could, he tells me, visualize their quarry, a large antelope with barley-twist horns called a kudu, from its tracks. He could see it tiring and slowing down, could feel the motion of the kudu's body within his own.

"It's a visceral experience. You feel you've *become* the kudu, and your own human body is following along somewhere behind. You're in and out of this trance-like, imaginative experience where you immerse yourself in the mind of the animal. Master trackers experience this for much longer periods and more often."

It is, he says, the ultimate condition of speculative tracking: where you feel you've *become* the animal.

Some scholars, Louis continues, believe shamanic trance states originated in persistence hunting. "When a tribal person performs a trance dance around a fire and has the spiritual, out-of-body experience of becoming a lion, what induces that trance state is the rhythmic clapping to which they dance until they're exhausted: the same physiological and mental elements that induce it in a persistence hunt."

Perhaps, says Louis, the half-man, half-animal therianthropes of ancient cave paintings actually portray master trackers becoming one with the animals they pursue.

Bernie Krause told me culture emerged out of the sounds of animals in the forest. Maybe art, science, and religion, I reflect, all have their earliest traces in those who tracked them.

These days, Louis still drives the same old Land Cruiser to track animals with the San deep inside the Kalahari.

"Why track?" he laughs, "Because yoga is boring!"

Over the last two million years, he says—99 percent of human history—we evolved to become highly specialized trackers.

"Now we've gone through the agricultural revolution, the Industrial Revolution, the information revolution. And the cognitive skills required for tracking are the one human aptitude that AI won't be able to replicate."

It's not, he concludes, just about following footprints. It's the art of scientific reasoning itself.

"Eventually, we'll all be trackers again," he tells me. "Just like in the beginning."

As June thrums on, I find myself less and less willing to go back inside. Outdoors, I slip into the present moment more easily: I empty my head and just *be*, sliding back and forth into Trust Technique "mindful regard" to observe the birds, the deer, the horses, without that harsh, human, Sedna-grabbing impulse. I empathize with flies stuck in the kitchen, carefully catch each one, and transport them outside. I open windows on chilly evenings until the kids begin complaining. Then I sit in the garden bundled up in a sweater till the coyotes start to howl.

All this listening, I realize, has made me reluctant to stop.

Midmonth, we sojourn in Puerto Rico. In San Juan at night, amid its tangle of streets, its grandiose old doorways, the crash of surf against its glowering fort, I'm looking elsewhere. At a ginger street cat, asleep at the base of a flight of mosaic stairs. She notices me noticing, stretches, and slinks over to say hello. At a quartet of junglefowl pullets, who leap cheeping from the roots of a tree and pluck up their collective courage to go pecking for scraps in the gutter: two, three, four, they plop down from the curb.

I've always loved to people watch; now I prefer tracking ghost ants across our breakfast table. High in the emerald mountains, I spot a pair of bright troupials cuckoo out from the knuckle of a coconut palm and pursue their contact calls—"I'm here!" "I'm there!" "I'm here!" "I'm there!"—all the way up the Rio Caonillas canyon. I attempt to deduce and induce: Where are they going? Why? What for? I lose them when I can't scramble as fast as they can fly and emerge at a huge, flat rock, past which the river cascades into a series of sparkling falls.

I'm about to turn back when a movement catches my eye.

I sit down. I stay quiet. And as time passes, I am no longer the disturbance. Around me reemerges a world of miniature dinosaurs—*Anolis cristatellus*, or crested anoles—each two inches long with a snubnosed snout and Gollum fingers. A foot away, one brave individual gives me a beady, cocked chicken eye. I divert my gaze. He scoots closer, closer, until he's on my foot, where, king of the castle, he inflates his bubble-gum dewlap.

The joy swells inside me. If I were a troupial, I'd sing it out. Instead, I keep quiet, and the anoles continue about their business.

A few days later, back home, I chat with Tamarack Song, founder of the Teaching Drum Outdoor School in the wilds of Wisconsin, who's been teaching wolf tracking and wilderness guiding for almost forty years. Tamarack, the very image of a wise old woodsman with his long gray beard and wire-rimmed spectacles, advises me not to attach myself to a sit spot at all.

"When you say, 'I'll sit here and watch,' you're being a passive receptor. We sit in front of a screen. We sit at a desk at school. We watch TV. It's inactive or reactive, but not interactive. A sit spot out in nature is still passive. Nature becomes that screen.

"Instead," he advises, "practice moving within nature, even in the city. Walking down a sidewalk from point A to point B, you could be thinking about who you're going to meet or what you're going to buy. Or you could try, like children do, not to step on the cracks. That will bring you simply into the now. While you're avoiding the cracks, you're noticing so much more. You're *there*."

I realize with a start that that's what I've been doing for months.

Every morning and evening, outside with the horses, I muck out their paddock. I fork hay. I sweep. I fill water. I pick out feet and groom out mud. I wander with them around the meadow, and I stand about, breathing, noticing, watching what they watch, seeing their stress indicators and subtle communications, being as much of an animal as I can.

I'm not staring with a twitcher's intensity or a hunter's adrenalized anticipation. I'm not begging for attention, not trying to tug back the wizard's curtain. I'm not even being silent: I'm just shoveling shit and humming country songs.

On June 19, a mouse ignores me and scurries over my right foot.

On June 21, Sven ignores me and runs by my left shoulder.

Barn swallows often sweep past the tip of my nose.

My pillow of silence, I understand, is deflating.

And there are knock-on effects too. Major is cuddlier. He uses his "go" button to ask to be stroked. Andvari lets me kiss his nose. The pigs accept me with open trotters—but then, they always do. I take the dogs out for ten-mile rambles, let them sniff and sniff, try to track what they're tracking. Sometimes time disappears. Wrapped up in the story of a footprint or a tree trunk scratched free from bark, I forget everything else. This is psychologist Mihaly Csikszentmihalyi's "flow state": moments of perfect focus and synchrony that he called "optimal experience." "It is what the sailor holding a tight course feels," he wrote, "when the wind whips through her hair, when the boat lunges through the waves like a colt."

And tracking is not just the purview of humans, I consider, as cows and deer pause their grazing to watch us pass by. Every living being is keeping track of someone else. Tracking is another animal Esperanto. I remember the day last summer when, sun blazing, we sweated and toiled setting fence posts around the meadow. Feeling that eyes-on-the-back-of-the-neck sensation, I stopped, looked around, and found my movements were being followed by two birds. The first: a female bald eagle high up in a birch, its crown bowing under her weight. The other was Mervyn, from the topmost twig of an old, gnarled cherry tree. One of the continent's largest and smallest birds, both sitting watching, tracking me.

On June 23, just after the solstice, three pigeons perch on the barn's spine in a summer rain: our happy couple and their one surviving youngster. The rain thickens. The parents retreat inside, but Baby, like me, stays out. Eventually, he flies away.

I imagine our starlings are relieved, after weeks of hard graft, to be empty nesters, but I wonder if the pigeon parents will be sad when their only child has gone for good. I'm projecting wildly, I know: Cairo will be flying the coop to university in September. Still, I've just read a scientific paper, playfully titled "Is Companion Animal Loss Cat-astrophic?," which cites a dozen studies, from whales to Western scrub jays, that "suggest that the psychological experience of loss may be widespread within the animal kingdom."

Carl Sagan's absence of evidence again, I think as I head indoors.

Later, the weather clears, and in a sky fit for astronomer Sagan, the fullest, brightest strawberry moon in eighteen years hoists itself over the valley. "A hunter's moon," says Indy. On a nighttime walk to admire it, we spot a jewel of a scarab beetle, resplendent atop a pyramid of horseshit. We follow as it descends and wobbles off like a clockwork toy.

Next morning, while the horses graze, I lie in the meadow reading two breathtaking reports from the far extremes of tracking. The first that scientists, tracking bacteria, have identified personalities in them. "Bacteria," says the paper, "appear to live lives of unforeseen behavioral complexity." The second describes how Kalahari trackers were recently able to identify over 90 percent of prehistoric rock-art animal prints in the Namibian mountains: not just the species, but also the sex, age, and which leg of the animal the print came from.

Really, though—I think of tracker Louis Liebenberg—they're not such extremes at all.

I pull out my book, *Soul Hunters* by anthropologist Rane Willerslev. Indigenous Siberian Yukaghir hunters, he says, become the elk while hunting it: "He was not an elk, and yet he was also not *not* an elk," Willerslev writes. "He was occupying a strange place in between human and nonhuman identities." Some people, I conclude, live with the insulation stripped off life's electrical wire. The natural world buzzes and fizzes in ways in which we, in our comfy modern lives, are inured to. Maybe that's why, aside from the dreaded horse-murdering glacier, I didn't mind hauling hay last winter in forty-belows. At least I could *feel* that fizzing.

I'm about to commence reading when I feel someone tracking me again.

I look up.

A brown bird sits in a blackthorn bush, too big to be a pigeon, too small to be a hawk. What is it? I forget myself, watch back, and it spooks her. She glides off, six feet above the tall grass.

It's a young pheasant.

An orphan? I hope not.

11

The Medium Is the Message

How to Listen in Pictures

"If a man does not keep pace with his companions, perhaps it is
because he hears a different drummer. Let him step to the music
he hears, however measured or far away."

—HENRY THOREAU, *Walden*

"In fact, what is any revelation or discovery but seeing things as
they are?"

—WILLIAM J. LONG, *How Animals Talk*

JULY

Earlier this year, around January, something funny started happening.

Twice on my way down to the meadow, I felt a sudden, fleeting
cramp in my abdomen and arrived at the corral to find Major with
an upset stomach. (How can you tell this in a horse? Lots and lots
and *lots* of farting.) One morning in March, I limped down for the
morning feed, my left toes smarting inexplicably inside my boot, to
discover Major had an abscess in his left hind hoof and was holding
it up in pain.

Hmm, I thought on both occasions. That's all. Just: *Hmm*.

And then, on a Sunday in May, I was stricken in the meadow with a
terrible tooth pain that moved from one tooth to the other, from upper
jaw to lower, until it filled my mouth. I could feel my jaw joints, or

TMJs, aching, as if I'd chewed too long on gum. It was instantaneous, horrible, and so intense it made me catch my breath.

I had never had a toothache before.

Ten minutes later, it was gone.

The next day, Dr. Dave—our local vet—arrived to check on Major's weight-gaining progress, which has been steady but slow.

"When were his teeth last floated?" he asked.

Horses' teeth keep growing throughout their lives. An annual float, filing away sharp edges and hooks that cause cuts and ulcers, ensures efficient chewing for extracting maximum nutrition from all that hay.

Nine months ago, I answered, at that boarding barn from which he came home limping and horribly skinny.

Dave nodded and checked his teeth anyway. "That must hurt," was all he said and set to work.

Hmm. I thought of that weird toothache. *Hmm.*

Since childhood, my emotions have always leaked out through my British exterior as bodily sensations. Doubled-over stomach aches that depart once the stress I've been ignoring does too. Anger that flames up my back when I witness someone being rude or unfair. A house guest who long overstayed his welcome gave me heart palpitations. It seems it's my body's way of yelling, "Hello? You up there. Can you deal with this please?"

But this is something new.

Although…not really. Sometimes when I'm with Andvari, I am suddenly flooded with anxiety, as I was on that rainy day with Trigger. Now I've learned to check myself. *This isn't mine,* I say. *It's his.* I remember Dr. Suzie Fay's dominant brain waves. I take a deep breath. Tune my inner radio to my own positive emotion. And feel him entraining.

Is it possible, I wonder, that since I've started listening more closely, I sometimes pick up on their physical feelings too?

I'm not a scientist. I'm not a professional trainer. I'm also not very spiritual. I believe this go-around on Earth is all we get, and we ought to make it a good one. Confusingly, though, I've occasionally caught a glimpse of something…else. This, I've found, characterizes many people's relationship to the "supernatural" ("forces or events," advises the *Cambridge Dictionary*, "that cannot be explained by science"). We label ourselves unbelievers—except for those bits that don't quite fit.

A woman tearing down my alphabet frieze one night when I was tiny. The teenage Ouija board session when all the lights went out. A figure in the window of the old house next door, soon after its owner passed away. Once, reporting at the Cassadaga Spiritualist Camp in Florida, where almost everyone's a medium or a seer, I woke in the night to the rocking chair creaking back and forth in my hotel room. I can explain all these away. A dream. A coincidence. A trick of the light. A drafty window. But then came the time with a toddler and a ginger cat.

Growing up, my grandpa was the only other person I knew who loved what he called a "good oss." He told me stories of riding to market with the farm's flock of sheep. He sat belting away at the piano as we sang "Horsey, horsey, don't you stop; Just let your feet go clippety-clop…" "Look! Oss!" he'd cry, when we drove off to visit relatives' farms. He idolized my sister and me. He would, he said, do anything for us. He died suddenly from a heart attack when I was twenty-six and living overseas.

Four years later, I was busy in our kitchen three thousand miles from England while Cassidy, our toddler, stacked blocks on the floor, and Denver, our cat, licked his paws in the sunshine. I was making coffee, chopping onions, thinking how my grandpa would have adored this blond, beaming being, when behind me, I felt someone about to touch my shoulder. Presuming Gal was home early, I turned to greet him. No one was there. But both cat and toddler had stopped what they were doing and were tracking something, slowly, intently, and in perfect unison, five feet above our kitchen floor. I watched them watch as whatever it was moved from behind me, rounded the corner, circled back, passed slowly behind me again, then up toward the ceiling and was gone.

The toddler returned to his toys and the cat to his toilette, and I burst into tears, somehow certain my grandpa had just visited us in the kitchen.

As a journalist, I've tried hard not to hold fixed beliefs. So I have never believed or *not* believed in psychics and mediums and animal communication: things I know nothing about. Still, certain images have always sprung to mind. Damask drapes. Chintz and patchouli. Flickering lamps. A scarf fringed with coins. I push through a curtain, cross a shadowy figure's palm with silver, hear a warbling "Is there anybody there?" There's a pause. A crackle in the ether. Then: "Is that you, Puddles?"

As many, though, swear by animal communicators as dismiss them. On July 2, my friend Ellen sends me notes from her communicator's recent "conversation" with her seven-year-old equine-assisted learning horse, Henry.

"He shows me a few different images of big red rubber balls (apples?)," the communicator writes, "and how he is trying to grab something with his teeth and that people are laughing and he's cheering them up. I don't know. Do horses play with balls?"

I read this, stunned. I know—but this animal communicator does not—that a huge, red, bouncy, blow-up yoga ball is Henry's favorite toy. Ellen has never shown this on social media. There's no way the communicator, who lives on a different continent, could know. Does—and *how* does—this work?

There are more things in heaven and earth, Horatio, I think, *than are dreamed of in your philosophy…*

I recall something Louis Liebenberg told me when he talked of shape-shifting and becoming the animal. "It's a kind of arrogance," he said, "not at least to be agnostic."

Everyone's favorite ethologist Marc Bekoff, it turns out, feels the same about animal communication.

"I'm open-minded about it," he says, surprising me when I cautiously mention it. "I'm the only scientist who would endorse Penelope Smith's book." Penelope—I look her up—has since the '70s been one of the world's leading animal communicators, credited with popularizing the field in mainstream America. "There's nothing to be lost," Marc continues, "by keeping the door open. I've been pretty impressed with a few animal communicators intuiting certain things without ever having seen the dog in question."

His policy, he says, is "'I don't know enough about it, so let's see what happens.' Isn't that what science is?"

On July 11, Indy digs out his pile of summer reading. *Swallows and Amazons*. A Sherlock Holmes anthology. And with them, *Your Psychic Pet*, a twenty-year-old book I once picked up for a dollar and have left in a packing case ever since.

He carries his stack across the road to our neighbor's farm, where he's taken a two-week job as pet sitter. His charges are Cooper, an exuberant golden retriever, and Nova, an indifferent long-haired tabby cat. Indy reads in the yard. He plays hours of fetch with Cooper. He hangs out with Nova on their living-room couch.

Now he's amusing himself a different way. From across the road, I receive a series of voice memos.

"Hi. I've decided to do some experiments from this psychic book," the first one says. "This one's called the 'What would you like to do?' test."

He outlines the instructions. Sit quietly, then mentally ask a dog that question. Often, author Richard Webster says, the dog will immediately appear, and you'll receive a mental impression of what they would like to do.

"So...er... It kind of worked with Cooper," comes the morning's next memo. "Well, he didn't actually *come* to me, so I had to go and find him. But when I did, he nudged his teddy bear, the one without the stuffing that we play with in the yard. So I'm guessing he wants to play outside. I didn't really need telepathy," he reflects, "to figure that one out."

Later, I receive another.

"Now I'll try the 'Come to me' test. You try summoning your cat without talking to them by sitting quietly and imagining them coming to you. 'It is unlikely you'll have to do this for more than five minutes,' the book says. So let's try this. I'll be calling Nova with my mind. Unfortunately, though," Indy adds, "I don't think this speaker picks up telepathic communication, so it'll just be quiet while I do it. OK. Go."

Silence for five and a half minutes.

"Um, it hasn't worked so far..." his voice returns. "I mean, she can probably hear me telepathically—she's just been staring at me this whole time with her ears pricked up. But I guess either Nova doesn't like me, or she just doesn't want to. This concludes the 'Come to me' test."

Throughout the afternoon, I receive further communications as Indy tries the "Find it" test. The "Food bowl" test. "Sending thoughts." ("I'll be thinking about Nova coming to eat and see if she comes over. Here we go." Three minutes later: "She might just not be hungry now. I'll try something else.")

At six, he returns for dinner.

"Weren't you skeptical?" I inquire over spaghetti.

"I had doubts." Indy shrugs. "But I have doubts about lots of things. Scientific stuff too. Doesn't mean I'm right. Maybe," he adds, "I just need practice."

I watch him cross the road, book in hand, back to Nova and Cooper to perform further experiments as logical to him, apparently, as testing for personality in earwigs.

In the early 1970s, not long before Irene Pepperberg walked into a pet shop to purchase the future Avian Language Experiment, a young woman who had never had doubts that animals could talk to humans was receiving "proof" of it from a small, traumatized black-and-white cat named Peaches.

Over fifty years later, famed animal communicator Penelope Smith is now a sprightly seventy-seven-year-old. She lives in Arizona, where she and Pepita, her Chihuahua, enjoy visiting their neighbor's rescued burros, Annie Oakley and Calamity Jane. Since those early days, she's taught thousands of people the principles of telepathic communication with animals, including celebrity psychics whose own courses run to hundreds, sometimes thousands, of dollars.

Born in 1946 in Chicago, "as a child," she tells me, "I always knew how to be quiet so as to communicate with small city animals: squirrels, birds, insects." She learned early, though, to keep this to herself. "People made fun of me. My mother constantly told me I was exaggerating, had a vivid imagination, even that I was lying."

Both her parents were alcoholics. "They were seeking in a bottle what they couldn't find in life." Penelope did her own searching. She spent nine months in a Catholic convent. She studied psychology and sociology. "Eventually I became a counselor, helping people through trauma."

Then, in 1971, someone left Peaches at her apartment. "She, like my human clients, was extremely traumatized and would only come near me. I decided to help her using the same counseling techniques I used with people, only telepathically."

Within a very short time, Penelope tells me, Peaches completely changed.

"She became so friendly that my roommates thought I'd switched cats. And I realized: wow. Not only could I still communicate with animals as I had as a child, but I could also help make their lives better. So I started doing so on the side. I wasn't charging anyone; I was just seeing what would happen. The results were phenomenal."

Throughout the '70s, she says, word spread. "People called, asking for consultations. I started traveling to their homes. I lectured. I did media appearances."

In those days, however, skepticism was rife.

"I'd be asked to go on TV shows just so people could ridicule or entrap me. I'm a strong person, but I'm sensitive too, and I'd vomit in the bathroom afterward. But for the most part, I knew what they were up to and would just bust through and get my message across. It was live programming, so they couldn't cut it out afterward." She grins. "I didn't let anything stop me."

Likewise in the printed press. "Many journalists were really good people, but their editors wanted a juicy headline. But let me tell you, even if they mocked me, it would still reach people. It planted a seed. And then when pet owners were desperate, when the vets and behaviorists couldn't help, they would seek out an animal communicator."

In the late 1970s, Penelope began teaching others and has been doing so ever since.

"When people start communicating with their own pets, with their little parakeet or pet fish, their whole world expands."

"That's wonderful," I say. "But *how do you do it?*"

"It's all about being quiet."

This, I think, sounds familiar.

"Our society promotes this tremendous anxiety. We're conditioned to have stimuli all the time. So first you must learn to come down and be quiet. Once you're quiet enough, anyone can do it: can open themselves up to their intuition."

So get present, she says. Be open. Be calm. "Tune in" to your animal. Some communicators, she adds, require physical proximity. Some need a photograph. Still others converse with the animal over distance with no visual aid at all.

"Then animals, who communicate on an energetic level, will share their state of being."

Some experience these "sharings," Penelope says, as pictures: a series of mental images that might show, for example, how an animal hurt itself. Sometimes it might appear as an exact verbal phrase. And some people experience feelings: pain or anxiety.

Hmm, I think. *Hmm*.

"It's not, though," she tells me, "some *Star Trek* sort of blinding lightning-bolt connection. Rather just gentle impressions. A knowing. A feeling. Which you then interpret, as best you can, with your human words."

"Then how," I ask, "can you differentiate between a communication and your own imagination?"

"Aha!" Penelope laughs. "That's the main question, the one every animal communicator has to answer. First thing: You need to ease all the thinking before you open up to hearing another. That way, if impressions suddenly arise, you will know they're not your own. Second, when you receive a communication from an animal, it's often a way of thinking you've never had. My dog will say things to me that make me say, 'Oh, I've never looked at it that way.'"

Also, she adds, if she receives a communication from an animal about a problem and thereby helps solve it, "that's how you know it's working."

I remember Major's teeth. And though I wasn't planning on it, I tell Penelope about the physical feelings I've been getting.

"I think," I can't believe I'm saying this out loud, "they might be coming from my horse."

Though I'm all alone when it happens, I admit, it feels as if there were someone on my shoulder, scoffing, "You're *imagining* it, you idiot."

It sounds ridiculous, I know. I brace myself to be mocked or pitied or, at best, told to take her course.

"Those are just the ideas you've absorbed," Penelope replies. "When you allow yourself to lean into this kind of listening, this little voice inside you will say, 'No, this doesn't exist,' 'What Penelope said is a bunch of hooey,' 'You don't have this,' 'She doesn't have it,' 'Everyone's lying.' And on it will go. Just observe that voice, shine your light on it, and say, 'Fine, but I'm going to do it anyway.' Then all that conditioning begins to lose its power.

"Beginners," she adds, "will get things in bits and snatches. I tell them not to share that with anyone. To put the fear of judgment or ridicule aside for now, get quiet, and see what happens. My job is to let people know this is nothing magical. It's a natural ability you've always had, which you can reclaim."

Over her life, she says, she's met many "old-timers" who are exceptionally good with animals. "You say to them, 'You're very telepathic,' and they'll say, 'Oh no, not me.' But they're getting their thoughts, their feelings, their intentions. I say, 'Well, it may sound woo-woo to you, but that's telepathy.'"

Nowadays, Penelope is teaching a course for children. "Children communicate naturally," she says, "which is why animals gravitate toward them. But we lose it as we age and become social animals seeking approval from our peers. Kids come in humility. They don't second-guess themselves."

This reminds me of an afternoon with my twelve-year-old nephew, Tamlyn, watching the signal crayfish who inhabit the banks around his English narrowboat home. "This one-clawed one," he pointed one out, "his favorite food is sweet potato. And these two here are good friends.

When I caught the bolder one in the net, the shy one came out and attacked it: I think he was trying to rescue him. Lots of these crayfish have friends and live together in next-door holes."

He tells me this as if it were the normalest, naturalest thing in the world—which, to Tamlyn and Indy, both young, scientifically minded naturalists, it is.

"And as for the doubters?," I ask Penelope.

She shrugs. "If it's too much for people, that's fine. We can just talk about the weather or the local grocery store."

Still, things have changed a lot, she says, in the last fifty years.

"Science is a tool; it's a way to observe. It's not a be-all and end-all. Things scientists refuted back then are now considered fact. Animals don't think, they said, they don't feel; their behavior is all automatic. No one says that anymore." (*Almost* no one, I think ruefully.) "I am," she concludes, "a very practical person. I go by results. And the fact is: This animal had a problem, and after we communicated, that problem resolved. As far as explaining it intellectually to a skeptic's satisfaction, I probably can't. And I'm OK with that."

Over the next few days, I dig a little deeper. I join animal communicator Allecia Maine and four of her students, each in a different physical location, in a session of their weekly remote "quantum science laboratory." They check in with a herd of rescued mustangs at the Eagles and Wild Horses Ranch in Colorado to see if the horses have something to say.

Group member Debi mostly receives their communications through feelings, she tells me, though "once in a while," she hears words.

Dolores, meanwhile, has learned "to trust my gut, not just my brain. I've reclaimed what I already knew, but was afraid to express. Where has it been all this time?, I keep wondering."

"It's all about being open," contributes Allecia. "But people are so unwilling to trust anything that's a feeling rather than a thought. They fear having made it up. They fear ridicule. And that feeling goes through so many filters that it's almost lost."

Her advice, she says, is to "conduct your own experiments, because you'll know instinctually if something is true. Ask yourself: Can I keep listening and let the world be what it is? Can I be as neutral as possible when there's a fight between my head and my gut feeling? Make yourself 'catchable' by the animal. Be the question."

This reminds me of something Tamarack told me. "A Blackfoot elder once explained to me: 'Tamarack, when you're talking, you're repeating what you already know. When you ask a question, that question is based on your prior knowledge and limits what you're going to receive. But if you *be* as a question, you're just open.'"

I consider what it means to be the question, to inhabit it, to feel around inside it, rather than simply demanding an answer. "What can I hear?" instead of "How could this work?" "What does this feel like?" rather than "Is this real?" And it reminds me of when my children come to me with their troubles. They don't want a reaction or a judgment; they don't even usually require a solution. They just want a space, a concave container, within which to be heard.

Next I connect with Diana Delmonte, an animal communicator and Zen meditation practitioner in Los Angeles, who was, she tells me, a "wholehearted skeptic" until her mid-forties, when her cat Bubby, she says, suddenly "spoke" to her.

Diana doesn't, she explains, need to be in the animal's presence to listen to them; helping find lost pets was her specialty for twenty years. Instead, she focuses her "thoughts and intentions" on the animal with whom she hopes to connect. "Get quiet," she says. "Get grounded. Focus on the breath. Get into your body and out of your head." Sometimes, she adds, she suggests practitioners imagine a golden light extending from their heart to the animal's.

"Animals think in pictures, and so do we," she tells me. "You can't recall a memory or think of a friend without seeing them in your mind. One aspect of animal communication is becoming aware of those pictures."

"But how then," I ask, "can you receive very specific messages—say, the location of a lost pet?"

"I pick up what I call a 'thought ball,' accompanied by the feelings and emotions the animal is trying to convey. I then interpret those thoughts and feelings into human words.

Not all communications, she adds, are messages the animal's owner wants to hear.

"I once did three communication sessions with a minihorse who was very angry and wouldn't let anyone near him. The first session, he said, 'Where's my mother?' The second, 'What happened to my mother?' The third, 'I want to know where my mother is.' And that's all I got."

Diana relayed this message to the horse's new owner.

"Oh no," the owner said.

The horse, it transpired, had been purchased along with his surrogate "mother," a goat the owner had subsequently sold on.

"Of course, when she heard what he'd told me, she was heartbroken. She never could get that goat back, so eventually she gave the horse away to a farm with other goats in the hope that he'd find happiness. I don't know," Diana adds, "if he ever did."

Her advice for aspiring animal communicators is similar to Penelope's: "In Zen meditation, boredom comes, and you let it be. Anger comes, and you let it be. That way, they dissipate. Thoughts are enticing. They come into your mind, but you don't have to think them."

Diana's Manx cat jumps up onto the desk and rubs herself against her computer camera.

"That's so weird," Diana says. "She never does this. Five years I've had her, and she's never done this before."

In mid-July, flash floods send gravel driveways on their annual tumble onto the county's roads. In a bucket, I measure seventeen inches of rain in just over a week. A trio of great blue herons flaps like pterodactyls over our flooded fields. I'm surprised to hear one silent sentinel *Craaaarck* at the others.

Heat makes the farm feel monsoon-tropical. Mervyn's raspberry patch has ripened overnight. I forget to bring my boots inside, and they're full of earwigs come morning. I gently shake them out and watch the Mabels dash for cover while the Ritas take their time.

As usual, Andvari doesn't mind the weather, but Major does. In fact, he seems to mind most things. He won't raise his front hoof for me to pick it. There are no more "go" signs to be groomed or petted. He nips at my sleeve. There's a tightening of his eye, an almost imperceptible clamping of his jaw. Signs I once wouldn't have noticed, but now I hear right away.

I resolve to keep listening till I work out what he's trying to say.

Meanwhile, I read up on quantum entanglement, something animal communicators often evoke to explain how their communication actually works and that Einstein called "spooky action at a distance."

It boggles my poor unscientific brain. But at its simplest, I find out, it's the law in physics whereby two particles that have been part of the same system and then move apart remain "non-locally connected"—entangled—so that a change in one affects a change in the other.

This concept was demonstrated in a 1997 experiment by Swiss physicist Nicolas Gisin: He separated two photons, or light particles, by a distance of seven miles and sent each in the opposite direction down an optical fiber. When one of the pair hit a two-way mirror, particle detectors recorded whether it went through it or bounced off. Whichever it did, its twin, seven miles away, did the same.

"Everything is energy. Everything is connected," Allecia Maine told me. "That one statement, with those two things, is quantum entanglement. On one level, we are separate individuals. On another, we are not. As humans, we want to separate out and give a label to everything, so entanglement is hard for our brains to understand. But it's not an intellectual understanding. It's an experience. You have to go into your 'gut brain' that feels the interconnectedness of everything."

Controversial Cambridge biochemistry professor-turned-paranormal researcher Rupert Sheldrake also uses quantum entanglement as an analogy for telepathy. Sheldrake performed his own tests: on dogs who "know" when their owners are coming home; on people who "know" when someone's about to phone them. (95 percent of women, Sheldrake says, experience thinking of a person just before they call. The men most sensitive to this phenomenon, he found, were in Argentina; the least—no surprise to me—were in the UK.)

In 2000, he also tested Aimee Morgana and N'kisi, the African gray parrot with the terrific sense of humor. N'kisi sat in one room and Aimee in another with one hundred sealed envelopes containing pictures of items N'kisi knew. One by one, Aimee opened the envelopes and thought of the picture each contained. In the first room, cameras captured N'kisi's vocalizations. The occasions of N'kisi saying what Aimee was seeing came in "significantly above chance."

"Rupert and I would both consider our research 'science,'" Aimee tells me when I ask her about it, "because we conduct it in a rigorous way, using classic scientific techniques. But it could also be a bridge between science and spirit, since we're studying something relating both to the concept of the flock mind and quantum phenomena of nonlocality and entanglement—which can refer back to older spiritual wisdom about how we are all connected."

Next I check in with my friend Kathy Price, a sheep farmer and energy worker in Mid Wales.

Kathy has an infectious laugh, a scientific education, and a farmer's down-to-earth sensibility. "I swear like a trooper," she tells me, "which can slightly spoil the spiritual effect. But working with sheep for thirty years,"—she grins—"is enough to make anyone swear."

Twenty years ago, she tells me, she attended a horsemanship clinic that "cracked her open." "People ignore their instincts and feelings, I realized, because society beats it out of you. Animals are shouting out loud at us, and we're saying, 'What was that? I can't quite hear you.' From that moment on, my paradigm shifted from a purely scientific lens to I didn't know the hell what but had to find out."

Kathy went home and read up on quantum physics.* "Everything is energy, I discovered. Everything is connected." She echoes Allecia. "In the quantum field, there's no separation."

Nowadays, aside from farm duties, Kathy works as a facilitator in energetic healing. "You don't need to believe it. You don't need to understand it. People say to me, 'Oh, it's a placebo.' I don't give a shit. All I know is that people get healed. I don't care what they call it."

Kathy's the second person I tell about the toothache, the foot ache, and the stomach pains.

"You should talk to my friend Cat," she says.

So I do.

Catriona MacDonald is a storyteller and animal communicator who lives on Cardigan Bay in Wales, a country Roger Deakin, in *Waterlog*, described as "stiff with magic." It is here that Cat, as she instructs me to call her, tells stories from the medieval Welsh epic, the *Mabinogion*, "on the lands on which they were written."

---

* Kathy's favorite introductory books on the subject, she tells me, are *The Field* by Lynn McTaggart, *The Biology of Belief* by Bruce Lipton, and *The Dancing Wu Li Masters* by Gary Zukov. I add them to my reading list.

Although she was trained in osteopathy, it was while transitioning to working on animals, she tells me, that she first heard one talk.

"I performed a session on a horse, and as I was walking away, I got a sudden, terrible headache. I had this sense that it wasn't my headache, but his."

This sounds familiar. I swallow and say nothing.

"I went back to look, and sure enough, he had a big lump on his forehead from a knock to the head. As soon as I realized that, my headache disappeared. And my question became how do I make this 'hearing' happen again?"

She was, she says, "remarkably slow at understanding the prerequisites."

First, she studied animal body language. "But I found that when I slowed down and cultivated stillness beforehand, there'd be a wave of feeling that came *before* the body language. Noticing the soles of my feet on the ground was important to do. Noticing the other sounds around me instead of homing in. Allowing things from my peripheral vision to slowly come into focus. When you settle in like that, stillness becomes the furniture and wallpaper of your internal environment. Then, when you're receiving a communication or a 'knowing,' it either arrives *into* that stillness or arises *out of it*. There's an ease to it. Trying too hard," she says, "will disturb that stillness, like ripples on a pond."

"How do communications arrive with you now?" I ask.

"I don't hear words or see pictures. What I experience are waves of feeling. Sometimes I'll smell or taste things. I communicated with a dog whose owner had just moved to the country. The dog told me, 'I can smell rabbits,' by me smelling them too.

It is, she tells me, a lot like tracking. "You're just in this place of listening. You let all the information you've been given settle, and then you wait for something to arise, mixing that information with the intelligence that's always there and is just part of life."

Cat invites me to join one of her Imaginal horse circles.

Not knowing what they are or what to expect, I accept.

I am sitting in a Zoom gathering with ten women from all over the world. We're a mixed bag: from Wales, Scotland, South Dakota, Denmark, Canada, New Zealand, Johannesburg. One, Catriona tells us, is a shamanic practitioner, here to keep the circle sacred. "Today," she continues, "we'll go on a journey to meet the Horse Nation—the spiritual presence of all horses who ever have and ever will live on Earth." Tears well instantly in my eyes. I have no idea why.

We introduce ourselves. Cat guides us through a "heart-centered" meditation. We break for tea.

When we reconvene, it's time, Cat says, to visit the Horse Nation, after which we'll divide into breakout groups to share the messages the horses gave us.

Oh shit, I think. It's my walk-like-a-coyote all over again. What if nothing happens? What if I don't believe in any of this? What will I tell my group then?

I picture Marc and Louis. "Why not try?" they'd say. "What harm can it do?" *Be open*, I tell myself, *and just see.*

So I close my eyes, take a deep breath, and, as the shaman drums us in, try to quell my self-consciousness, my self-judgment, the feeling that I probably look quite silly.

And the second I do, I startle. Because there they are, before my closed eyes: a herd of horses on a rise in the far distance. It seems they've been waiting to say hello.

It is not a hallucination. I've taken no ayahuasca or mug of mushroom tea, not even so much as a cold lunchtime Chardonnay. Nor is it a dream; I'm totally awake. Is this, I wonder, what people mean by a vision? It's clear: like a movie, like…real life. I can't influence what they're doing either, can't disrupt the picture with some strange, silly detail—ice creams appearing on trees, the French president on a llama—just to prove I'm still in control.

The herd comes galloping down to greet me, two horses leading the way. To the left: a big, bold dapple gray. To the right: a calmer, quieter bay mare. They step forward and touch their noses to my chest.

Immediately, a light beams out of it and traces the shape of a white luna moth.

What the *hell*?

The image changes briefly, the rail trail at the bottom of our meadow, and I'm galloping on a horse I don't know. No fear of speed, of falling, just joy and exhilaration.

Back to the herd, and two final things. In front stands little Mars, a miniature horse I knew who died a month ago. He's happy, content. And just beneath my line of vision, a small soul hopping up and down, trying hard to be heard. It's a pony, or a foal maybe. A pinto, patchy brown-and-white. *Wait for me*, he's saying. *Wait for me*.

"And now," I hear Cat's voice from far away, "time to come back. Thank the Horse Nation, and start feeling your body, your fingers, your toes…"

I blink my eyes open. I've no idea how long that all took. Two minutes? Twenty? I jot down notes, but there's no way I'm forgetting this. Despite all my Britishness and misgivings, it was just too…*real*. I feel as though I were emerging, squinting, from Plato's cave, marveling at all the things outside my experience that were there just out of sight all along.

We break into groups, and I find myself with two kind ladies: one from rural New Zealand, the other from Montana.

Frowning, I try to explain what I've just seen.

"I think that's my mare," says the American when I reach the bit about the calm bay horse. "She does that."

"Mm," echoes the New Zealander. "Lovely. So lovely."

"Carry what you've received," Cat advises us all as the meeting ends, "like a treasure in your pocket."

Later, I turn to Google. *Symbolism luna moth*, I type.

"Seeing a luna moth can be a sign to embrace change and appreciate the natural world around you."

Luna moths "signify new beginnings…a continuing quest for truth and knowledge, the gift of intuition, psychic perception and heightened awareness. Things that are associated with 'seeing the light.'"

Luna moths "are symbols of transformation and rebirth… You're moving on from the person you used to be and becoming a more spiritually enlightened version of yourself."

I see, I think and pour a stiff gin and tonic. Because Roger Deakin, it appears, was right.

Next morning, very early, grass spiders have slung a thousand glistening hammocks across the orchard. I shower, then walk, towel-wrapped and dripping, into the kitchen.

I sit down in the window seat and think about these slippings of the veil, these peeks across the frontier. I glance at the picture, still pinned to the kitchen wall, of Polish naturalist Simona Kossak and her enormous boar, Żabka. "I crossed the threshold and found myself on the side of the trees and animals," she said. "I therefore speak on their behalf."

I am starting to know what she meant.

Outside, down flutters a sharp-shinned hawk.

I gasp. These exquisite, reclusive birds, *Accipiter striatus*, the smallest North American hawk, hunt in low woodland, pouncing on unwitting songbirds. I've never seen one in real life before, let alone up close. Yet here he is, this xanthous, tweedy fellow, with his umber breast and dabbed yellow beak.

He hops up to the kitchen window, casts one circumspect eye in my direction. Desperately, I let go my racing thoughts: of writing this down, of how long he'll stay, of getting a photo, of how it would be to tame one like Naomi in England. I let them all drip from me like the water puddling on the kitchen floor.

He looks at me. I look at him looking at me. He looks at me looking at him looking at me. And as with Sedna, only stronger, another of those spaces appears, those rents in the fabric of the everyday. Timeless. Serene. We're part of the same everything.

And then he's gone, off into the wood. And I'm back inside human me, dripping wet in a bath towel.

Two turkey vultures, I notice, are circling above the house: someone else's pillow of silence.

I make tea, get dressed, and go out to pick raspberries for breakfast. One cane springs back and whips me in the face. I hear someone giggle.

Was that...the *raspberry canes*?

Then I notice the tag on my tea bag.

"Talking to plants," it quotes herbalist Rosemary Gladstar, "is one way of talking directly to Spirit."

What the *fuck*? What is happening to me?

"Every little pine needle," Thoreau wrote in *Walden*, "expanded and swelled with sympathy, and befriended me." He didn't say they *giggled* at him, though.

I don't tell my family. I hesitate, even now, to write this down. Instead, I go inside and email Catriona. Do I sound, I ask, like a lunatic?

"I love what you experienced," Cat writes back. "My first workshops were with trees and plants: amazing beings with as much to say as the creatures. It doesn't surprise me that you had an experience like this as you open your channels of listening. And no," she adds kindly, "I don't think you're a lunatic."

It's appropriate that I meet Dr. M. J. Barrett at a botanical garden, since she first came to animal communication via listening to her plants. It is a bright, hot Monday, and we walk a woodland path, our faces pushing through spiderweb streamers. There's a power-line hum of cicadas. Jays call overhead. We settle on a bench in a delicious patch of shade.

M. J. heads the Department of Intuitive Interspecies Communication at the University of Saskatchewan, one of the world's only universities studying animal communication in its many forms.

"IIC," the department's website says, "presents as a detailed, non-verbal and non-physical form of communication between humans and other animals. Drawing on a diversity of intuitive capacities, IIC includes the mutual exchange of visceral feelings, emotions, mental impressions and thoughts, embodied sensations of touch, smell, taste, sound, as well as visuals in the mind's eye. While these exchanges can occur while in direct physical proximity to the animal, they can also occur over great distances and without the need for visual, auditory, olfactory, voice or other cues that humans normally associate with direct interactive communication."

I'm amazed to see what I've been striving to understand described so scientifically.

M. J. was, she tells me, working on a PhD in outdoor education when she met a group of women "who were communicating with the land, using pendulums and dowsing." One day, she mentioned to one of them that some potted plants in her garden were not thriving and asked if she knew why.

"Why don't we ask them?" the woman replied.

"I didn't resist," says M. J. "I was excited and intensely curious. I couldn't not explore it."

The plants "told" her, via a technique known as pendulum

divination, that they'd prefer a sunnier spot. "I realized their roots had been cold."[†]

Soon after, she says, she switched PhDs to begin exploring interspecies communication. "Animal communicators in diverse cultures," she tells me, "are consistently and intentionally able to connect and engage in communication of some kind. Exactly what happens in those communications we still don't know. And apparently, they get results. Our research group is not about assessing or proving those results. It's about understanding the lived experience of animal communication."

The IIC department investigates this lived experience, she says, via phenomenology.

"In simple terms," explains a 2019 Uniformed Services University report, "phenomenology can be defined as an approach to research that seeks to describe the essence of a phenomenon by exploring it from the perspective of those who have experienced it.... Phenomenology is the study of an individual's lived experience of the world."

I love this, I tell M. J., because the communicators I've met have all emphasized it: *This is my experience*, they say. *Take it or leave it. I'm not trying to buy you in.* I think of all the people, from Pamela Clark to Naomi Johns to Warwick Schiller to Alexandra Horowitz, who put the individual animal's lived experience first and wonder if there's such a thing as interspecies zoophenomenology. Nagel or not, I decide, there should be.

"Different knowledge systems," M. J. continues, "have different epistemologies, or ways of knowing."

For example, she explains, for many Indigenous peoples, interspecies communication practices are part of everyday life. Sydney Kuppenbender, M. J.'s graduate student and member of the Métis First Nation, writes that not all Métis and Cree tales she heard as a

---

[†] Pendulum divination, or dowsing, uses a swinging pendulum to divine the answers to yes/no questions.

child, of tricksters and other ancestors coexisting with animals, were metaphorical. "My ancestors could in fact," she says, "communicate with animals." The tricksters of First Nations and Native American stories are shape-shifters, taking forms such as Coyote, Blue Jay, and Raven. "Métis stories are not 'make-believe,'" reiterates *The Indigenous Peoples Atlas of Canada*. "They are an integral component of the Métis worldview. The Elders or Old People, moreover, believe them to be true."‡

But some cultures, says M. J., don't recognize other cultures' methods as valid. "And that's a problem. For instance, if anecdotal evidence or alternative ways of knowing have no value in Western science, that's epistemological ignorance."

Fortunately, she says, the University of Saskatchewan has a very strong Indigenous orientation and great respect for and valuing of Indigenous knowledge."

Thanks to this, the IIC department has performed community-engaged research with First Nations elders and knowledge keepers. Meanwhile, Sydney is examining case studies of Indigenous animal communication with wildlife. What value, she wants to know, can this knowledge bring to Western science?

M. J. tells me this is an expanding field.

"Ten years ago, you had to belabor your point. Now there's a groundswell of scholars moving the dialogue forward in terms of the legitimacy, validity, and importance of other ways of knowing. Still," she says, "it's vital to be aware of the sacredness of many Indigenous animal communication protocols. There are blurry lines. Some people are more willing or able to share than others."

And this varies immensely, she explains, from tribe to tribe.

"The Anishinaabe, for example, aren't supposed to talk about what

‡   The 2006 book *The World We Used to Live In* by the late Native American scholar and activist Vine Deloria Jr. provides a breathtaking glimpse into this topic.

happens in the sweat lodge. The Cree, however, can share more things. Fortunately, there are incredible Indigenous scholars who can do this work."

As with animals' capacities and communications, I reflect, just because *we* can't see something doesn't mean it's not there.

In the gardens, our pool of shade has receded. We get up to leave.

"Different ways of knowing are valid and legitimate," M. J. concludes, "and there are lots of scientists who talk about the role of intuition in their breakthroughs. It just often gets written out. But I learned a long time ago that you're not going to reach everybody. You just do good science with ethical intent."

Later, cooling down in a café, I read that the IIC department "challenges deeply inscribed Cartesian dualisms by disestablishing absolutist understandings of the dichotomies of body/mind and nature/culture."

Bloody Descartes again, I think. If ever I tell someone about the raspberries and they ridicule me, this is what I'll say I was doing.

By late July, there is still no sign of Siegfried, but the sun, at least, has returned.

In the meadow, Indy and I find a pupa hanging on the fence. We can make out the shape of the morphing caterpillar inside. Will it keep its caterpillar memories? Which moth will it become? We're not sure. I think of the author Rachel Carson, whom M. J. quoted in the sultry botanical gardens. "'I sincerely believe that for the child,'" said M. J., "'and for the parent seeking to guide him, it is not half so important to know as to feel.'"

Back up at the house, we see Mervyn sharing the feeder, for the first time ever, with two other hummers.

"I think they're his children," says Indy.

It's his last day petsitting Nova and Cooper, and he's off to try more telepathy. "When a young child tells you they had a conversation with a bee," M. J. also said to me, "it's quite possible they did."

The word *clairvoyance* comes from the French, meaning "clear sight." Though I tried to curb it, a part of me expected smoke and mirrors. Instead I found generosity and open-mindedness. Like tracking, these communicators don't consider their work apart from science, rather a part of science that hasn't been much explored.

Have I dismantled that frontier between us and animals? Not yet. But I've slipped through from time to time, as if into Narnia. In my recent lived experience, I've met the fathomless gaze of an octopus. I've sensed the multitudes contained in chimpanzees, witnessed the wonder of a trusted merlin in free flight. I have shared space with green monkeys and green anoles. I've felt the incomparable biology of joy with one red horse, two orange pigs, and one yellow pony. I've found what M. J. calls "interspecies humility" and "a conversation between equals" and come to understand that that frontier is one *we've* erected, not nature: a fata morgana we can dissolve, if only temporarily, just by forgetting it's there at all.

I have also discovered, I reflect as I watch Indy cross the road, that our ways of knowing are no more the only ways than our ways of seeing those yellow lines on asphalt are the only ways of seeing. "We see things not as they are but as we are," wrote George T. W. Patrick in his 1890 essay, "The Psychology of Prejudice," "that is, we see the world not as it is, but as molded by the individual peculiarities of our minds." Perhaps there's dichromatic knowing. Tetrachromatic knowing. A million other shimmering shades of knowledge we simply do not know about.

My head spins at this breathtaking thought.

I go inside and sit down to read.

First a report, almost twenty years old now, in the *New England Journal of Medicine* by Dr. David Dosa, detailing the extraordinary life of Oscar the cat, who correctly predicted more than fifty deaths in his Rhode Island nursing home's advanced dementia unit. "His mere presence at the bedside," wrote Dosa, "is viewed by physicians and nursing home staff as an almost absolute indicator of impending death, allowing staff members to adequately notify families." Nobody knew how Oscar did it.

Then stories describing lost dogs who find their way home, sometimes over thousands of miles. Are cryptochromes responsible? Or magnetoception? (A Czech study found dogs prefer aligning themselves along the earth's north–south axis magnetic field to poop; to ascertain this, researchers studied 1893 discrete, or perhaps indiscreet, poops.) As with Oscar, nobody knows.

I return to the meadow, where Andvari comes up to say hello. He's gone from being too afraid to let me touch him to soliciting kisses on his buttermilk nose. But it took us eighteen months to get there.

Some days now, when I ask if he wants to go out for a ride, "Yes," he says. "Let's go."

Other days: "No thank you" or "Not likely."

There's no wrong answer.

Major, on the other hand, still doesn't seem right.

So this morning, I do all the things. First, a body scan. Then I still my mind. I breathe. I feel my toes. I try to be the question. I doubt myself and try not to.

Zoophenomenologically, I let my scratchy radio tune itself in.

In no words, no images, just one powerful wave of uncomfortable feeling, I get my answer.

It's not me. It's him.

12

An Unfamiliar Feeling

How to Listen When You Don't Want to Hear

> "And of course there must be something wrong
> In wanting to silence any song."
>
> —ROBERT FROST, "A Minor Bird"

AUGUST

High, dry summer. Our neighbor's peach tree droops with fruit. The year is a concertina being squeezed: 10:00 p.m. dusk becomes 9:15 becomes the end of the 8s. Apples ripen. The orchard's old trunks, heartwood gone, are still producing and producing—a bumper crop this year. Some long-departed farmer, in a fit of whimsy, grafted three varieties onto one particular tree. One limb bears red, one russet, and one green.

Constance and Agnes are now a solid fourteen months old. They keep cool in matching paddling pools.

"The majority of pigs," says the British Kunekune Society, "will reach finished size around 12 to 15 months old."

Just like earwigs, I think: the subtle power of language.

I slip it into conversation with my teenagers. "Well," I say when they compare heights, "you've probably reached finished size by now."

They don't like this turn of phrase. "Dehumanizing," says Cairo.

"And anyway," adds Zeyah, "eating pigs is inhumane."

Humane.

I look it up. "Marked by compassion, sympathy, or consideration for humans or animals."

That's what listening can do. It makes humans humane. And there's never been a better time for it; about 150 species, big and small, are vanishing from the planet every day. Many will leave before we ever hear them or know they were here at all.

When I was a little girl, I had a persistent thought: How do I know, I wondered, that I am just *I* and not *we*? That I'm not sharing my me-ness, or at least part of it, with other creatures and am simply not aware of it? This thought was ephemeral, elusive. As soon as I tried to grab it, it was gone. And it always left me with one image: of looking through the eyes of a deer in a light-dappled wood, me but not me, both I and the deer unaware we were actually part of each other.

In brief blips this year, I've felt it again—that I'm me, but also us. A feeling so intense it's almost supernatural—yet equally not at all. I suspect this is how animals feel all the time.

But it's a double-edged sword. Because as August bakes, I feel more of Major's aches and pains. I notice the little things. I hear him better.

Despite long hours of summer pasture, his lovely chestnut bum that we worked so hard to make fat again is shrinking. He seems stiff. Sometimes he drags his back feet. A subtle lameness shifts from leg to leg. He still won't pick up that front foot. I research, add supplements, stretches, painkillers. I develop a constant low-level stomachache, which gets worse the closer I am to him. I remind myself it isn't mine. I wonder if it's his stomach muscles, compensating for something wrong in his legs, his hips, his spine.

Still, he's eating well. He whinnies his singular, deflating-cow moo when he sees me coming with an apple. Some days, he seems better, and my human mind tells me I'm imagining it all.

Then one morning as I'm shoveling shit, Major leaves his breakfast, walks over, and to my astonishment rests his forehead against mine, exactly how he did the day my grandma died.

Only this time, the message is different.

It's not words. It's not sounds or smells or pictures. Just a clear feeling…a knowing. An intense weariness that says: *I've had enough.*

No, no, my mind struggles. *You're only fourteen—barely middle-aged. We can try more injections. More X-rays. Other exercises or painkillers. We can gallop those trails again.*

Thanks, says that feeling. *But no thanks.*

I do not want to hear this, my mind yells. *I don't.* It feels like the ultimate test of listening: hearing someone tell you they need to go when you want them to stay. It's hard not to dismiss it as nonsense. Not to cover my eyes like Loulis the chimpanzee the day the workers came. But I wasn't expecting this. It wasn't even a half-formed thought in my head. And that's how I know it came from him.

I remember that promise I made in the car, furious, in March in the thrumming rain.

I'll always listen, I said. *Even if it makes things uncomfortable.*

So I stand, and I do, and it does.

Later, I call the vet. I book an appointment. I play it down.

"Nothing urgent," I tell Wanda, the receptionist. "He's just a little off. A lameness exam, that's all."

Both vets, she says, are very busy. "Three weeks' today? Is that OK?"

"That's fine," I answer too cheerily. "No urgency, like I said."

For the next two weeks, I feel nothing. I hear nothing. I'm numb, except for this nagging, indefinable pain.

The cows across the river become garden ornaments.

Birdsong is elevator music.

Andvari prefers Indy. Wildlife won't come close.

The curtains have swished shut on the magic show.

Why? What am I doing wrong? I miss it, and it makes me anxious. I feel like binoculars, trying to focus. The dogs hover. Everyone else stays away.

One evening, I sit outside and reread Henry Beston's 1928 *The Outermost House*, one of my favorites.

"We patronize them for their incompleteness, for their tragic fate of having taken form so far below ourselves. And therein we err, and greatly err. For the animal shall not be measured by man. In a world older and more complete than ours they move finished and complete, gifted with extensions of the senses we have lost or never attained, living by voices we shall never hear. They are not brethren, they are not underlings: They are other nations."

It just makes me feel worse.

I shuffle through the stupefying summer days, clumsy, incongruent. I feed animals. I do tasks. I knock things over. I can't keep my mind on mindfulness. Sometimes my heart skips anxious beats, or I feel a flare of panic in my stomach. I'm claustrophobic, like when I was six years old and stuck in a broken elevator. I'm panicked, like when I was ten and stuck on a seized-up ghost train. But mostly I feel nothing. I lose my appetite. I fall over a gate, tear off a toenail, and bruise my knees.

Driving up our lane, a little bird flies into my windshield and dies instantly. I pull over, walk back to find it: a perfect warbler in sulfur and Prussian blue, its tiny talons curled.

I feel nothing. I hiccup in sorrow. Then feel nothing again.

I cup the warbler in one palm, spark extinguished, and realize something. I'm shutting things out because it's just too much. I don't want to be the one who listened and, because I did, had to make a decision

I don't want to make. So I'm stuck in my nervous system, somewhere between sympathetic flight and dorsal freeze. Freeze is there for our protection. It's to stop us feeling too much. Feeling nothing is better than feeling everything. A worm baking in the sun. This little bird. A big red horse, slip-sliding away.

I suddenly understand why I felt nothing, years ago, in that hellish chicken farm.

"Incoherence," Kathy Price told me, "is survival mode. You can only perceive matter; you can't feel its energy. It's fear, lack, limitation, in place of love, light, and abundance."

That's it. That's exactly how I feel. A splinter sheared off from the interconnected world.

It might seem a win-win-win to listen to animals: for humans, for animals, for the planet. But listening holds you accountable. If you don't listen, you can claim you simply never heard.

I have a veterinarian friend who says she's often caught between listening to her patients and to the wishes of their human owners. Depression and suicide rates among vets are shocking; American veterinarians are three to five times more likely to die by suicide than the general population. Mary Lee Jensvold has written papers on the toll of "compassion fatigue," the "traumatization of helpers through their efforts at helping others," in Fauna's chimpanzee caretakers.

Suddenly, at lunchtime one day, I remember something I've blocked out for years: the time when, for work, I had to visit a desert livestock market. Someone there was slaughtering terrified geese, as casually as if they were shelling peas: grab one, stuff it upside down into a metal cone, slice off its head, let it bleed out, keep up the market small talk. I trudged through a Western-style shopping mall later that day, half-numb, half-incredulous that some lives could go on unwitting, eating burgers and buying fur-trimmed parkas and feather pillows, while others were being extinguished.

My feet felt leaden; my mind and spirit matched. It's better, I thought, to feel nothing than to feel all this. Freeze explains the things we do to animals and to other humans. You shut your ears to shut your heart.

On day 15, I can stand it no longer. I need someone else to listen, to see if they hear the same thing. Because—that panicked pang again— what if I'm wrong?

So I email Catriona.

"Could you communicate with Major for me, please?" I ask.

I send her no details, no red bouncy yoga balls. All she needs is his name, though I also email her a headshot photograph of him three winters ago, looking noble in the snow.

"I've never done this before," I tell her.

She "talks" with him early next morning, long before I get up, sitting quietly in her cottage in Wales, three thousand physical miles away.

Then at 7:00 a.m., as dawn bleeds in above the farm, Cat and I meet on Zoom.

"When I communicate with an animal," she says, "first I sit down and establish our 'heart connection.' Then I just wait for whatever they want to bring into that meeting place. I don't ask questions. I just sit and be."

Be the question, I think. Already my eyes are pricking with tears.

Major came in gently, she tells me, but with physical noise all around him, "like the sonic boom of an aircraft when it splits the sky."

His body aches, she reads from her notes of their conversation. He showed her a tremendous weakness in his hind quarters. His hips. His spine. One front leg in particular. His soul feels youthful, Cat says, but his body feels damaged, too old for its years, too exhausted to even find rest. His heart might be failing too: Twice she felt small palpitations. He fears the coming winter, when the cold will feel like daggers.

"He loves you," she says, "and he wants to be free."

This, I understand, is where story is finally useful to listening. He was worked too hard, too early. Someone "rode the frigging hair off him." That's why, at fourteen, his body is that of a horse twice his age.

As I listen, sorrow squeezes my sternum till I can't draw breath. My throat closes up, an anaphylaxis of grief. I try to speak but only gasp.

"Thank you," I manage.

Then I spring a leak.

I start to cry. And find I cannot stop.

That night, I stand with Major, my face buried in his neck, and cry and cry and cry. I tell him he's a good boy and that I love him, that it's all right, and that I know.

For the first time in weeks, my radio tuned in hard without the static, I feel it all, and it feels good, and it feels terrible.

On day 21, Nicole the vet comes.

"I don't think it's just a lameness exam anymore," I tell her.

We watch Major walk. He looks worse. He won't—can't—trot. She examines him and tells me all the same things Catriona did. His hips. His back. That front leg. Perhaps his heart.

We talk through the options.

"There's still things you could try," she says, "but you have to think who you'd be trying them for."

I shake my head. That grief again, strangling my words.

Better to let him go before the cold and damp, she says, on a good day, loved and with a full belly.

We make an appointment to euthanize him in two weeks' time.

I cry a lot these next two weeks, but my listening's returned, and the world floods back in.

One morning, I tear down that picture of Simona Kossak and her beloved boar. She started me out on this journey; now I'm angry with her. If I'd never heard of her, perhaps I would never have heard my horse. I could have gone on in blissful ignorance, believing everything was fine.

A year ago, I longed to listen like she could. Now, *be careful what you wish for*, I think regretfully.

My anger subsides, and I flip through the notes I made on her, back when I gathered every English scrap I could find. "I finished my studies in biology," Simona wrote, "but it wasn't until my years spent in the forest that I learned to understand the language of animals. And I know it so well now that I should be burned at the stake as a witch."

Witches, I consider, have familiars. The word says it all.

M. J. told me there are people talking with animals on every continent except Antarctica. I decide to connect with a couple more.

First Manjiri Latey, an animal communicator from Pune, India, on the edge of the shimmering Western Ghats. She is the daughter of mountaineers; her mother studied under Tenzing Norgay, one of the first two people to summit Everest.

Manjiri is an expert on India's tribal and Indigenous peoples' relationships to the natural world. "There's an old saying," she says. "'A body creates an illusion of separation from nature.' But many Indian tribal peoples see themselves as branches of one tree, whereas we'd consider ourselves each a separate tree, forgetting that the network beneath connects the roots."

She also runs animal communication workshops for "pet parents, the curious, confirmed skeptics. And," she adds, "for army gentlemen guarding India's northern borders, who are already using their intuitive abilities for survival: asking animals for guidance, asking mountains whether anyone's burrowed in or hidden arms or ammunition."

She has, she tells me, conversed with a forest, an atom, an ocean, hives of wild bees.

"In India, in our apartment complexes, bees sometimes make hives where there used to be trees. Typically, the complex will bring in a company to burn the hive down."

But sometimes, she says, the human residents will ask her to come in and speak with the bees first.

"In many cases, you're speaking to the collective, interconnected whole, not to the queen bee, for example. 'Are you aware,' I'll ask the hive, 'that this apartment complex sees you as a menace?'"

"Yes," she tells me hives will answer, "but there's no other place to go."

"They are ideating on burning the hive down. What is your take on this?"

"Can you ask them to give us a week, to find somewhere to move?"

Then, Manjiri tells me, she will speak with whoever requested the consultation.

"They might say no. If they do, I go back to the bees. 'They said no.' I'll say, 'Do you have any other ideas?' And the bees will come up with wonderful, creative solutions. Like: 'Could the apartment dwellers keep their windows shut between X and X times of day while we go out to find alternative places?' So then there's a negotiation. Rarely do I, the mediator, need to generate an idea. I simply frame the questions."

Be the question, I think again.

"Usually," Manjiri concludes, "some solution is reached, and the bees leave."

Tribes in India, she tells me, enjoy close mutual listening with local wildlife. "In Central India, I went out one day with tribal trackers employed by the forest department in a forest inhabited by an aggressive tigress with four cubs. I was in a jeep, but these guys were on foot, and I said to them, 'You're on foot, knowing this tigress might have her cubs on one side of you and her kill on the other. She's even been known to charge vehicles. What's going on in your minds?'"

"We already informed her," they replied, "that we're coming for her good."

"They were so matter-of-fact," Manjiri recalls, "in their shorts and their flip-flops."

Rounding a corner, they encountered the tigress, lying in the middle of the trail.

From the jeep, Manjiri held her breath and watched.

"She simply got up," Manjiri says, "and moved to one side of the trail. The trackers took the other. It was seamless."

It's a joy to be able to hear again.

Andvari lets me back in, and we play with obstacles, absorbed in the flow. When I make a mistake and his reins get hooked on a pole, he drops his head and waits for me to save us. Being listened to has given him trust and courage.

Despite our best intentions, stories linger. You can't take away what's been done to you or anyone else. You can't fully unlearn an old language. Animals with a troubled background might never have the optimism of a Constance. But neural pathways can get brambly and overgrown. We can become palimpsests: the old stuff faded and overwritten.

From the city, Tyger sends me videos. The girls have bought more cat trees. Winnie sits in whichever's in the sunshine, deliciously happy, scowling at strangers. I work on CAT with the dogs. Both improve with summer visitors, though Puff slips the leash and chases the young man here to start construction of a new, unglacial corral. "She's friendly! She's friendly!" I scream like a madwoman as Puff attempts to dog-flea herself into his arms.

The pigs' harnesses arrive, and we commence leash training. Indy wants to take them to the farmers' market next summer. We plan the incremental steps to get there: climbing a ramp, sitting nicely in the car.

"To market, to market, to walk a fat pig," I butcher a rhyme my grandpa used to recite. "Home again, home again, jiggety-jig."

My friend's Muscovy ducks have ducklings. I promise Indy three for his birthday.

"Huey, Dewey, and Louie," he says.

"Noam Chomsky, Herbert Terrace, B. F. Skinner," I counter.

We listen to a podcast for kids in which linguistics expert Dr. Jing Huang says the first human words were imitations of animals' grunts, hoots, and cries.

Your nurturing instincts will expand, reads my fortune in a cookie, *to include many people.*

I decide "people" means "chickens": some throwaway ladies someone else doesn't want whose conversations will brighten my life. We might even adopt a turkey for Thanksgiving.

Another friend's mare births a foal: a little chestnut pinto.

Hmm, I think. Just: *Hmm.*

I'm sitting working in the study one day when a movement at the window makes me look up.

It's Mervyn, hovering outside.

I've never seen him around here before, on the opposite side of the house. There are no flowers left on the lilacs. The hydrangeas are already going dry. I made sugar water earlier, I remember, but forgot to fill the feeder up. Could it be he's come to find...*me?*

When I reach the kitchen, Mervyn's already back at his usual window, waiting.

I replenish his feeder with trembling hands.

Meanwhile, I commence the hunt for a flock of Jacob sheep: a whole field full of creatures we'll name after mythical horned animals like Pan—from Arcadia, where it was probably never windy—and Mr. Tumnus, who always carried an umbrella.

Gal expresses an interest in llamas. He retracts it quickly when he sees me go for my phone.

And I spend lots of time with Major. He aches. Doesn't want to be groomed or kissed or touched. So we stand side by side in parallel inactivity as swallows swoop over the meadow. I feel my toes inside my socks. Hear birds all around. Notice the rhythm of my breathing. And I don't vanish. I just…mingle. With the trees, the lives, the lichen, with the wind and birds and insects, and I feel that feeling I had as a child, when I wondered how I knew I wasn't also a deer in the forest.

One afternoon, as I sit outside, the hummers play a game of chase around my head. I feel like a Looney Tunes character, dazed, with bluebirds. I am surrounded by a living, *BRRRR-ZZZZZZ-UMMMMMM*ing Calder mobile. That night, with the Milky Way whorling above us, I'm lying outside with Indy, listening to coyotes harmonize in the woods, when something the size of my hand lands on the farmhouse wall.

A white luna moth.

Next morning, I'm out feeding the horses when I see something furry on the road. One of our young chipmunks, clipped by a car, lying dead in the middle of a lane. I bite my lip. Poor thing. Then I see him breathe.

Oh no.

I run up and look again. He's on his side, chest heaving. I look both ways: no traffic…yet. What to do? I could pick him up with my bare hands, but what about rabies?

No time either to run for gloves. If another car comes this way, the little guy's a goner. I grab two fallen branches from the ground. "I'm sorry," I say, "I'm sorry, sorry, sorry." And wincing, I roly-poly him to the verge.

No blood. But he's contorted like bad taxidermy because of how I've poked and prodded him. I race into the house, grab those gloves, and move him gently from verge to beneath a bush. I lay him out more comfortably. Perhaps he's in freeze, I think.

I wish he could tell me how he feels.

I become aware of sounds around me. Chipmunks' squeaks telegraph out from the cherry tree, from the trees across the street. They fusillade from the trees throughout the wood. It sounds like a dozen, maybe more.

Are they warning him or one another? Do they think me the predator?

I spin around slowly. I can't see a single one. "It's OK," I say. "I'm trying to help. I'm sorry. I'm really sorry."

I go out on the school run. Return. He's still alive but hasn't moved. Internal bleeding? Broken limbs? I no longer have the confidence of a thirteen-year-old with a pigeon.

"What shall we do?" I beseech Gal.

He postpones his morning's work, finds a cardboard box, and takes our little patient on an eighty-mile trip to the nearest wildlife hospital.

I can't work. I can't think. I feel sick. Did I hurt him, knocking him off the road like that? Am I responsible for his death?

Two hours later, Gal calls. "The vet saw him, and he's going to be all right. They'll release him in a week or two. You saved his life."

I'm so happy I can't stop crying. Like a madman, I go outside and tell the chipmunks and the birds and the raspberries.

Though Mervyn and family, I notice, have flown.

Kerri Lake in Arroyo Grande, California, is a warm, generous presence. Five minutes into our call, I feel we've been friends for years. I have heard from Warwick about her animal communication work. But on her website, one phrase leaps out at me: "Recover the gifts you knew as a child."

"My earliest memories," she tells me, "are of lying in the grass, looking at clover flowers, at the insects buzzing all around, and it all occurring to me like music. This music, I knew, would still exist even if my house, my yard, weren't here. It was so clear to me that all of life was communicating harmonically that I assumed everyone else was hearing it too."

It took her years to realize that her family, within which there was frequent conflict, "was not hearing the music."

These days, Kerri teaches people "how to reopen themselves to that musical, energetic world of harmony and interconnection. A friend of mine calls it 'the hum.'"

"The hum," I repeat, relishing the onomatopoeia.

Kerri starts, she says, with one simple question. "Ask yourself: *What else is here?* Leave some space for it. The moment you entertain that question, you'll be presented with something new in your awareness. People will say they're not intuitive, that they don't feel anything. I say it's not that you don't feel; it's that your mind may not have a relationship with how to find it."

To help them find it, she says, she asks questions. When have you just *known*, she'll ask, that your dog needs to go out? Or when have you met up with a friend, and they tell you they're doing great, but you know they're not?

"And they'll say: Oh, *that's* communication?"

Animal communication, Kerri tells me, is simply an expansion of that capacity, developed to be aware of more nuance.

"It's zooming into a gut feeling until it's superclear, with lots of detail. It's not 'learn this because you don't know it'—it's 'recognize where you *do* already know it.' That helps debunk the mystery of it. It's about achieving harmony and in that harmony finding all the information we need.

From a survival perspective, she continues, confirming my suspicions, freeze or shutdown is beneficial.

"It lets you be in control, it keeps things predictable, and it creates a comfort zone. But it's not really living."

I agree: On reflection, I felt far worse in that "comfort zone" of freeze, more isolated, empty, incongruent, alone, than I do, however painful, when I am listening.

"Nature's not interested in comfort. When we're moving through life in survival, we're encapsulating ourselves as if we're not related to everything else. Instead organize yourself to *offer* connection instead of trying to *obtain* it."

Relax, she exhorts. Ask, "What else is here?"

"Create a space in which something can happen. Slow and easy. It's OK. The animal kingdom will always meet us wherever we're willing to be met."

After our call, I go outside. *What else is here?* I ask.

I hear ospreys cry churlish over the river.

I hear our starlings chatter.

A blue dasher dragonfly, Lalique-magnificent, whirrs in to land by my coffee cup.

The biophony, I think. The hum.

The night before it happens, there's a terrific storm. I wonder if that's why the hummers left so early. Thunder cracks. Lightning illuminates the valley in photographic negative, the river a quicksilver gash.

In the last few days, Major's pulled further away. His walking has worsened. It reminds me of my grandma in her final days of withdrawal from the world, before my big red horse stood with me in my raw, knotted grief.

I feed him a huge dinner while the storm rages and Andvari stands guard outside.

Next morning, I bring the horses breakfast. Major greets me with his deflating-air-mattress honk.

I'll never hear that again, I realize.

I feed him a bucketful of apples from the traffic-light tree.

Horses live in the present moment. He doesn't know this was his last breakfast. That he'll never compete in that interspecies soccer match.

But I do. That's our human curse. If I could talk to the animals like King Solomon, it's the one thing I wouldn't tell them.

In the early months, hanging out with a "difficult" big red horse, I'd sing "Graceland," and he'd yawn and sigh.

When Andvari was scared, I'd sing a Sanskrit chant my friend Amrita taught me. "Om mani padme hum…"

Now I turn to poetry.

A hawk skims the thermals, and I recite "The Windhover." "'My heart in hiding; stirred for a bird—the achieve of, the mastery of the thing!'"

A grasshopper lands, lucid, near my shoe, and I change to Keats.

> The Poetry of earth is never dead:
>> When all the birds are faint with the hot sun,
>> And hide in cooling trees, a voice will run
> From hedge to hedge about the new-mown
>> mead;
> That is the Grasshopper's—he takes the lead
>> In summer luxury.

Then Nicole the vet arrives.

She examines Major.

"You're doing the right thing," she says.

Together we lead him down the meadow to his favorite napping spot.

The sun is out. Orb weavers' webs span spikes of goldenrod: two-inch spiders, black-and-yellow with fangs. They are totally harmless. But *I'm scary*, they broadcast to anyone who doesn't know better.

A breeze plays through Major's mane.

"Are you ready?" asks Nicole.

The word *euthanasia* comes from the Greek for "a good death." Isn't that the ultimate privilege in this world—a good life, followed by a good death? Still, I feel the tug between my humanness, which feels sad and guilty and grieving, and something older, bigger, inside and all around me, that gently tells me it's OK, that everything is part of everything and there's no sorrow in this: It just *is*.

Tears well. I think of all the things we did together. The trails we rode. The trails he wouldn't. The sticks he once fetched in the snow. That last small ride that I didn't know would be. And of the things I'd hoped we'd still do. Of what I could have done differently or better.

And I nod, because my throat won't let me say it.

It's quiet and peaceful. He touches my hand with his nose one last time. "Go," he's saying. And then he's gone.

In this split second, I feel my own feelings. Grief. Nostalgia. Confounded hope. Helplessness. Compassion. Loss. Love.

But also, just as powerfully, I feel his.

Peace.

I walk up to the house alone, as energy still fizzing through Major's eyelids causes them to flutter as if he were dreaming. In one hand, I hold his lead rope and halter; in the other, a lock of his mane.

On the way, a tremendous flurry from the depths of Indy's pumpkin patch. Six pheasants explode upward and flap off into the meadow where they're subsumed, like my big red horse, by the meadowgrass and goldenrod.

Fräulein, Siegfried, Hildegard. And this year's crop of youngsters.

The days afterward are a fug of grief.

I expect my body to feel instantly better, but it doesn't. The aches subside slowly.

And there's one more poem, by Kipling though not about Solomon, that I just cannot shake. Snatches inhabit my thoughts day in, day out, until I pull out a book and read it outside, out loud.

> There is sorrow enough in the natural way
> From men and women to fill our day;
> And when we are certain of sorrow in store,
> Why do we always arrange for more?
> *Brothers and Sisters, I bid you beware*
> *Of giving your heart to a dog to tear.*

[...]

> When the body that lived at your single will,
> With its whimper of welcome, is stilled (how still!).
> When the spirit that answered your every mood
> Is gone—wherever it goes—for good,
> *You will discover how much you care,*
> *And will give your heart to a dog to tear.*

It's not just dogs. It's horses. Cats. Hamsters. Stick insects. Guinea pigs. Rabbits. Chipmunks run over by cars. Baby birds caught by the jaws of our loving cats. Why do we do it, again and again? Why do we long so much to listen? Because deep down, beneath the civilization, the conditioning, and the socializing, we still know it in our bones. We are animals too, they remind us. We all belong together.

On September 4, I sit in the sunshine on the kitchen doorstep. A late-season fawn, just outgrowing her Bambi spots, looks up from the orchard, then goes back to grazing.

"Happy we who can bask in this warm September sun," I've been reading Thoreau again, "which illumines all creatures, as well when they rest as when they toil."

I shed a tear. My nose starts running. Then I notice a jumping spider on a leaf beside me, black and silver and effulgent, as if embroidered into life. I look closer. Her little bank of eyes looks back. I move my head one way. Her body does the same.

My radio crackles.

Tell the story if you have to, said Nahshon Cook, then plant a flower and move on.

I walk down to the meadow and plant a weeping willow on Major's grave.

I Say Goodbye, and You Say Hello

> "The Setting Sun will always set me to rights, or if a Sparrow comes before my Window, I take part in its existence and pick about the gravel."
>
> —JOHN KEATS, November 22, 1817

LATE FALL

Summer gives way to fall, Keats's season of "mists and mellow fruitfulness." The baby pheasants grow up, strike out, disperse. I hope they stay safe with hunting season underway. One afternoon, mid-harvest, I discover a wasps' nest, a perfect paper lantern. I contemplate contacting Manjiri but instead decide to leave them to their own devices. Wasps pollinate orchards and meadows. They eat crop-damaging greenfly and blackfly. I tell them I'm grateful every time I walk by, just in case they can hear me.

In September, three children head off to university. Our pigeons settle into life as empty nesters. The starlings continue to expand their joyous repertoire, singing out the power tools of autumnal farm work.

For a while after Major's death, Andvari reverts to that anxious pony who can only say no. So I listen. I happy him up, let him know I value his opinion. I take in an equine lodger, a little chestnut three-year-old to keep him company.

"Would you like to go for a ride?" I ask Andvari one morning.

"OK," he replies at last.

So we do, and I can feel he feels things will be all right.

Soon after, I start finding poop in the ponies' new corral that does not belong to them. Recalling my day at the beach, I stoop to examine it. This shit is canine, I determine, but can narrow it down no further. So Indy relocates the trail cam.

A red fox, we find, has been slipping in to join them at night. He brings toys: an apple, a dead mouse. He capers, tosses, catches. Sometimes he curls up and sleeps. Sometimes he poops. Three feet away, the ponies sit watching like proud parents at the playground.

Why does he come, night after night? Maybe it's for safety. Maybe he enjoys the feeling of sand beneath his paws. Or perhaps another interspecies friendship is flourishing in the midst of all my own.

In October, Dolores the groundhog packs her bags and departs for the old cow barn. Sven the squirrel defends an autumn acorn cache with barks and rattles and chattering teeth. A fat Eastern American toad takes up residence behind an upturned flowerpot, waddling out each evening to snag his supper. Howard and Penny, the loved-up Canada geese, return one weekend for a couple's retreat in the meadow. Siegfried, an outraged civil servant, voices his concerns.

On the first night of November, a coyote howls beneath our bedroom window, a wild, scrappy Romeo.

Life goes out, and life comes in, and it's all talking. In the skies. On the ground. In the soil that holds Major in its earthen embrace. What Thoreau calls an "infinite and unaccountable friendliness." Or what I, thanks to Kerri, now call *the hum*.

What did I learn, in the end, from my year of listening? I've asked that on soft autumn days as leaves fall and skies lower and rain drifts up the valley.

I've learned to trust my feelings. To get really, really, *really* quiet and listen with my eyes, my skin, my intuition, as much as with my brain and ears.

I've learned to feel less desperation, less anxiety, if I don't know what an animal is saying: to relax into it as if deciphering the nuances in a loved one's letter, rather than jumping up and down behind a high brick wall, yelling, "Hello? Hello? Is there anybody there?"

I've learned I'm more animal than I was before—or perhaps just as much as I was all along but simply had forgotten.

From box breathing to ninety-second emotions to fuck the story to keeping an eye on AI, I've gained tools with which to watch and wonder. I have also started reading scientific papers, for fun, on Saturday nights (take that, osmosis).

I've learned that just as every tiny being is its own feeling, speaking someone, we are all simultaneously part of a greater uncut cloth. Both Buddhists and Hindus talk, like Manjiri, of an "illusion of separateness," but perhaps surprisingly, Albert Einstein did too. "A human being is part of the whole," he wrote in a 1950 letter, "called by us 'Universe,' a part limited in time and space. He experiences himself, his thoughts and feelings as something separated from the rest—a kind of optical delusion of his consciousness."

I have learned that listening makes life both exquisite and unbearable, causing me sometimes to envy the Jain Buddhist monks I met in India, who, masked and barefoot, sweep the ground before them with soft brooms, "to avoid," says a 2013 article in *American Entomologist*, "treading on insects and other small organisms as this is seen as 'treading on souls.'" But I am not a Jain monk. I silence voices inadvertently as I go about my day. We ordinary humans still have to draw our limits, guard our hearts.

And I've learned I'm with Keats on negative capability. Because we'll never really know if what we, the receivers, are hearing is precisely what they, the senders, mean. I'll likely never find out what Mervyn Tugwell's true message was when he sent those wartime words. I've learned to see beauty in that, not frustration.

One of the first things I ever learned to do when interviewing people was to pause. They think they've finished their answer. You think they've finished their answer. And your instinct is to immediately fill the silence with something else: a clarification, the next question, admiration. But instead I learned to leave a little bubble of silence. A pause. Not a disconnection, but a resting place. From that resting place, the interviewee inevitably came up with the best thoughts of the entire interview.

I realize now that *that's* being the question. A gentle pause, a silence. The chance for the world to tell you something you didn't expect to hear. But there's something else too. I've dropped that scala naturae hierarchy, from angels down to rocks, that thanks to Aristotle, we can't help but carry around. I'm as happy to sit beside a spider as an elephant. I'm interested in what everything has to say. "Nothing is especially wonderful," wrote naturalist William Long, "when all is wonder." I will never vacuum a fruit fly up again.

How can you listen better to your own animals at home or in your environment?

I recommend picking your own unique path through science, training, and intuition. Select from each just the bits that appeal to you and to your animal. Get quiet, experiment, seek the shared joy, and don't worry about discarding the things you find don't suit.

Explore your edges. Push into realms of yourself you think are silly or feel uncomfortable. When you hear a voice talking, don't assume you are imagining it; look to science, ask your intuition, shrug your shoulders, and just see. Your animal won't judge you. They won't mock you or talk behind your back (or if they do, they'll only be saying how great you are). No one except your dog has to watch you walk like a coyote or hear you gobble like a turkey. You don't need tell anyone but your cat if you hear the raspberries giggle. You don't have to publish your results in a science journal or spill your tears in front of supportive strangers—though you could do all those things if you like.

And trust yourself. You know more than you think you do, you can hear more than you think you can, and it's thinking, in any case, that often gets in the way. Irene Pepperberg didn't know what would happen the day she purchased her Avian Language Experiment. Warwick Schiller wasn't sure if listening to damaged horses would do anything at all. Master trackers never know for certain they're on the right trail. You can't really get it wrong: All listening is good listening, even if, like Mary Lee Jensvold, you discover something that's uncomfortable to hear.

I stopped listening as closely to animals when I started caring what the human world thought of me. And though I don't wear kilts or turtlenecks and don't cut my own hair (usually), I now care less what people think. And I'm *British*. If I can do it, you can too.

Now, when life gets busy, when the human world intrudes and builds tension inside me like a pressure cooker, I stop. I feel my fingers. Feel my feet on the ground. I make a little space and ask, "What else is here? *Who* else is here?"

I'll wait. I'll watch. And there will always be something. A note of birdsong on a treeless city street. A hastening ant. An unhastening earwig. A simple act of listening, of science, of self-training, of connection and empathy that pulls me back out of my Cartesian head and into my animate, animal body. Every single time. Like the

Wart, the future King Arthur in *The Once and Future King*, I feel "uncreated."

The day Major died, I stood afterward with the local gravedigger, a soft-spoken man named Willy, who, with his wild beard and haunted eyes, couldn't look more Dickensian if he tried. I'd seen him waiting at a respectful distance, hat doffed, as I'd walked Major down to his meadow resting place.

Now, work done, he leaned against his tractor, smoking a hand-rolled cigarette.

When not digging graves, he told me, he manages hard-to-reach woodlands the old-fashioned way, hauling lumber out with animal power.

"Always had a team of oxen helpin' me," he said. "Secret to workin' oxen is you've gotta trust 'em."

"But how on earth," I said, "do you trust a two-thousand-pound ox?"

Willy sucked thoughtfully on his roll-up. "Listen to 'em, and they'll listen to you. Let 'em talk before you talk back. It's just conversation, isn't it? Simple as that."

Simple as that.

Acknowledgments

Thank you to the many experts who so generously let me badger them with questions. Thank you to Zeyah, Cairo, Tyger, and Cassidy for agreeing to get back to their muddy roots. Thank you to the indomitable Indio for being both chief entomologist and head ungulate keeper. Thank you to my glorious agent, Gillian MacKenzie, who as a child tried to will wild birds to talk to her. Thanks to Olivia Turner and Anna Michels at Sourcebooks for being brilliant, kind, and encouraging, and to Emily Proano, Patience Bramlett, Angela Corpus, Hannah Kil, and Emily Janakiram for their incredible support and hard work. Thank you to Fran, Ellen, Deb, Andrea, Lisi, Natalia, and Paula for listening to my every animal-listening woe, and to Tamlyn and Celandine for being ever-intrepid field researchers. And to Gal for pretending not to like animals, then fussing over them more than I do.

Notes

INTRODUCTION

"The destiny of Man": T. H. White, *The Once and Future King* (G. P. Putnam's Sons, 1958), 221.

"The method, then, is simply": Charles Foster, *Being a Beast* (Picador, 2017), 13.

a heart that drums: "Hummingbird Hearts Beat 10 Times Faster Than Yours," National Audubon Society, November 4, 2021, https://www.audubon.org/news /hummingbird-hearts-beat-10-times-faster-yours.

Hummingbirds are intensely territorial: "Why Do Hummingbirds Fight So Much?," All About Birds, Cornell Lab of Ornithology, accessed December 3, 2024, https://www.allaboutbirds.org/news/why-do-hummingbirds-fight-so-much/.

nests bound by spider's webs: "Ruby-Throated Hummingbird," Hinterland Who's Who, accessed November 27, 2024, https://www.hww.ca/en/wildlife /birds/ruby-throated-hummingbird.html.

"There is no more sagacious animal": Jules Verne, *A Journey to the Center of the Earth* (Ward, Lock, 1877).

"My entire sympathy": *Simona*, directed by Natalia Koryncka-Gruz (Eureka Studio, 2022), https://dafilms.com/film/15924-simona.

"tried to flee to the Soviet Union twice": Zbigniew Święch, "A Paradise Called Dziedzinka," *PRZEKRÓJ*, Issue 50–51, 1974, https://przekroj.pl/archiwum /artykuly/70252.

She filmed nature films about them: Janus R. Kowalczyk, "The Extraordinary Life of Simona Kossak," Culture.pl, July 22, 2015, https://culture.pl/en/article /the-extraordinary-life-of-simona-kossak; "Kossak Simona, prof. dr. hab. 1943– 2007," Encyklopedia Puszczy Bialowieskiej, accessed September 5, 2024, https:// www.encyklopedia.puszcza-bialowieska.eu/index.php?dzial=haslo&id=27; "Prof dr hab. SIMONA KOSSAK," Forest Research Institute (archived website), accessed September 4, 2024, https://web.archive.org/web/20070825165910 /http://www.ibles.waw.pl/struktura/zln/pracownicy/skossak.htm.

"There was never a king": Rudyard Kipling, *Just So Stories* (Doubleday Page, 1912), 249.

Bees exchange directions: Shihao Dong et al., "Social Signal Learning of the Waggle Dance in Honey Bees," *Science* 379, no. 6636 (March 2023): 1015–18, https://doi.org/10.1126/science.ade1702.

Elephants express love and grief: "Can Elephants Feel Emotions and Empathy?," World Animal Protection, last modified June 6, 2024, https://www .worldanimalprotection.org/latest/blogs/elephants-emotions-and-empathy/; "How Elephants Communicate," ElephantVoices, accessed December 3, 2024, https://www.elephantvoices.org/elephant-communication/why-how-and-what -elephants-communicate.html.

Dolphins each possess: Virginia Morell, "Dolphins Can Call Each Other, Not by Name, But by Whistle," *Science*, February 20, 2013, https://www.science.org /content/article/dolphins-can-call-each-other-not-name-whistle.

Mother cows and their calves: "Do You Speak Cow? Researchers Listen in on 'Conversations' Between Calves and Their Mothers," *ScienceDaily*, December 15, 2014, https://www.sciencedaily.com/releases/2014/12/141215101612.htm; Amanda Searby and Pierre Jouventin, "Mother-Lamb Acoustic Recognition in Sheep," *Proceedings of the Royal Society B* 270, no. 1526 (September 2003): 1765–71, https://doi.org/10.1098/rspb.2003.2442.

Bats, whales, and chickadees: Yosef Prat et al., "Crowd Vocal Learning Induces Vocal Dialects in Bats: Playback of Conspecifics Shapes Fundamental Frequency Usage by Pups," *PLOS Biology* 15, no. 10 (October 2017): e2002556, https:// doi.org/10.1371/journal.pbio.2002556; Taylor A. Hersh et al., "Evidence from Sperm Whale Clans of Symbolic Marking in Non-Human Cultures," *PNAS* 119, no. 37 (September 8, 2022): e2201692119, https://doi.org/10.1073/pnas .2201692119; Lori E. Miyasato and Myron C. Baker, "Black-Capped Chickadee Call Dialects along a Continuous Habitat Corridor," *Animal Behaviour* 57, no 6 (June 1999): 1311–18, https://doi.org/10.1006/anbe.1999.1109.

a 2018 scientific paper: Balint Z. Kacsoh, Julianna Bozler, and Giovanni Bosco, "Drosophila Species Learn Dialects Through Communal Living," *PLOS Genetics* 14, no 7 (July 2018): e1007430, https://doi.org/10.1371/journal.pgen .1007430.

bees' ability to recognize human faces: "Bees Recognize Human Faces Using Feature Recognition," *ScienceDaily*, February 8, 2010, https://www.sciencedaily .com/releases/2010/01/100129092010.htm.

"It is greed": "Democritus: Quotes," Britannica, accessed March 15, 2025, https://www.britannica.com/quotes/Democritus.

Anna Kaminska: Anna Kaminska, in email correspondence with the author, August 2023.

eight foul-mouthed parrots: Sarah Kuta, "These Parrots Won't Stop Swearing: Will They Learn to Behave—or Corrupt the Entire Flock?," *Smithsonian Magazine*, January 26, 2024, https://www.smithsonianmag.com/smart-news /how-this-zoo-is-handling-its-foul-mouthed-parrots-180983655/.

Hoover the orphaned Maine harbor seal: Patricia A. Currier, "Hoover Will Talk No More: A Delight to Thousands, Aquarium Seal Dies at Age 14," *Boston Globe,* July 26, 1985, https://www.proquest.com/docview/294275556?sourcetype =Newspapers.

"I should not talk so much": Henry David Thoreau, *Walden* (Walter Scott, 1888), 2.

CHAPTER 1: I SAY TALKING, AND YOU SAY SPEAKING

"The one great barrier": Friedrich Max Müller, *Lectures on the Science of Language: Delivered at the Royal Institution of Great Britain in April, May & June 1861*, vol. 1 (Longmans, Green, 1862), 360.

"The difference in mind": Charles Darwin, *The Descent of Man, and Selection in Relation to Sex*, vol. 1 (John Murray, 1871), 105.

"the imparting or exchanging": "communication," Dictionary, Google, accessed December 30, 2024, https://www.google.com/search?q=definition+communication.

"speak in order to give": "talk," Dictionary, Google, accessed December 24, 2024, https://www.google.com/search?q=definition+talk.

"say something in order": "speak," Dictionary, Google, accessed December 24, 2024, https://www.google.com/search?q=definition+speak.

"Say," it informs me: *Merriam-Webster Dictionary*, s.v. "say," accessed December 30, 2024, https://www.merriam-webster.com/dictionary/say.

animals do not dream: Hanoch Ben-Yami, "Logic and the Boundaries of Animal Mentality," in *Wittgenstein and Beyond: Essays in Honour of Hans-Johann Glock*, ed. Christoph C. Pfisterer, Nicole Rathgeb, and Eva Schmidt (Routledge, 2022), 243–53.

We read up: Shannon M. Digweed, "Allow Me to Introduce Myself: Squirrels Use Rattle Calls to Identify Themselves," *The Conversation*, April 11, 2022, https://theconversation.com/allow-me-to-introduce-myself-squirrels-use-rattle-calls-to-identify-themselves-175896; David R. Wilson et al., "Red Squirrels Use Territorial Vocalizations for Kin Discrimination," *Animal Behaviour* 107 (September 2015), 79–85, https://doi.org/10.1016/j.anbehav.2015.06.011.

hummingbirds have an exceptional memory: "A Hummingbird Never Forgets," *Science*, March 6, 2006, https://www.science.org/content/article/hummingbird-never-forgets.

wild crows recognize human voices: Claudia A. F. Wascher et al., "You Sound Familiar: Carrion Crows Can Differentiate Between the Calls of Known and

Unknown Heterospecifics," *Animal Cognition* 15, no. 5 (September 2012): 1015–19, https://doi.org/10.1007/s10071-012-0508-8.

researchers donned a rubber caveman mask: Heather N. Cornell, John M. Marzluff, and Shannon Pecorano, "Social Learning Spreads Knowledge about Dangerous Humans among American Crows," *Proceedings of the Royal Society* 279, no. 1728 (February 2012): 499–508, https://doi.org/10.1098/rspb.2011.0957.

a 2020 Japanese study: Sabrina Schalz and Ei-Ichi Azawa, "Language Discrimination by Large-Billed Crows" (paper presented at EvoLang 13, Brussels, Belgium, February 2020), http://dx.doi.org/10.17617/2.3190925.

anecdotal evidence of hummingbird-human interactions: r/birding, "Do hummingbirds beg?," Reddit, accessed December 9, 2024, https://www.reddit .com/r/birding/comments/18bofps/do_hummingbirds_beg/.

mankind's earliest known figurative cave paintings: Franco Viviani, "Finding and Losing the World's Oldest Cave Art in Sulawesi," *SAPIENS*, April 27, 2021, https://www.sapiens.org/archaeology/sulawesi-cave-paintings/; "Half Human Half Animal Rock Art," Bradshaw Foundation, February 12, 2015, https://www .bradshawfoundation.com/news/cave_art_paintings.php?id=Half-Human-Half -Animal-Rock-Art.

"The animals other than man": Aristotle, *Metaphysics* 1.1, https://classics.mit .edu/Aristotle/metaphysics.1.i.html.

Animals were regularly prosecuted: E. P. Evans, *The Criminal Prosecution and Capital Punishment of Animals* (William Heinemann, 1906), 150; Sonya Vatomsky, "When Societies Put Animals on Trial," JSTOR Daily, September 13, 2017, https://daily.jstor.org/when-societies-put-animals-on-trial/; Brett C. Ratcliffe, "Scarab Beetles in Human Culture," *Coleopterists Society Monograph* 5 (2006): 85–101, https://digitalcommons.unl.edu/entomologypapers/94.

"For it is highly deserving of remark": René Descartes, *Discourse on the Method of Rightly Conducting One's Reason and of Seeking Truth in the Sciences* (Leiden, 1637), https://www.gutenberg.org/files/59/59-h/59-h.htm.

"The lower animals": Charles Darwin, *The Descent of Man and Selection in Relation to Sex*, 2nd ed. (John Murray, 1874), 85–86.

"interpret an action as the outcome": C. Lloyd Morgan, *An Introduction to Comparative Psychology*, 2nd ed. (W. Scott, 1903), 53.

Austrian ethologist Konrad Lorenz: Konrad Lorenz, *King Solomon's Ring* (Methuen, 1961).

B. F. Skinner: Javier Virués-Ortega, "The Case Against B. F. Skinner 45 Years Later: An Encounter with N. Chomsky," *Behavior Analyst* 29, no. 2 (Fall 2006): 243–51, https://doi.org/10.1007/BF03392133.

Skinner boxed pigeons: Audrey Watters, "Pigeons, Operant Conditioning, and Social Control," *Hack Education*, June 15, 2018, https://hackeducation.com/2018/06/15/pigeons; "Behavior: Skinner's Utopia: Panacea, or Path to Hell?," *Time*, September 20, 1971, https://time.com/archive/6843845/behavior-skinners-utopia-panacea-or-path-to-hell/.

W. H. Thorpe boxed baby chaffinches: W. H. Thorpe, "The Learning of Song Patterns by Birds, With Especial Reference to the Song of the Chaffinch Fringilla Coelebs," *IBIS International Journal of Avian Science* 100, no. 4 (October 1958): 535–70, https://doi.org/10.1111/j.1474–919X.1958.tb07960.x.

Noam Chomsky: Noam Chomsky, "A Review of B. F. Skinner's 'Verbal Behavior,'" in *Readings in the Psychology of Language*, ed. Leon A. Jakobovits and Murray S. Miron (Prentice-Hall, 1967), 142–43, https://chomsky.info/1967____/.

"It's about as likely that an ape": Noam Chomsky, "On the Myth of Ape Language," interview by Matt Aames Cucchiaro, email correspondence, 2007–2008, https://chomsky.info/2007____/.

Skinner, in the best academic tradition: B. F. Skinner, "On Having a Poem," lecture given at the New York Poetry Center, New York City, October 13, 1971, B. F. Skinner Foundation, audio, https://www.bfskinner.org/bf-skinners-lecture-on-having-a-poem/.

Newsweek **published a letter:** Skinner, "On Having a Poem."

Others put more birds: Michael J. Rorbaugh, Wesley C Zaynor, and Joseph H. Grosslight, "Reinforcement Control of 'Autistic' Verbalization in the Mynah Bird," *Psychonomic Science* 11, no. 3 (March 1968): 91–92, http://dx.doi.org/10.3758/BF03328149.

Harvard zoologist Donald Griffin: Donald R. Griffin, *Animal Minds: Beyond Cognition to Consciousness* (University of Chicago Press, 1992), 34–35.

seven primary emotions: Jaak Panksepp, "Affective Neuroscience of the Emotional BrainMind: Evolutionary Perspectives and Implications for Understanding Depression," *Dialogues in Clinical Neuroscience* 12, no. 4 (2010): 533–45, https://doi.org/10.31887/DCNS.2010.12.4/jpanksepp.

discovery of the *FoxP2* gene: Kat Fowler, "Simon Edward Fisher (1970–)," Embryo Project Encyclopedia, April 27, 2017, https://embryo.asu.edu/pages /simon-edward-fisher-1970.

found in rodents and fish too: D. M. Webb and J. Zhang, "*FoxP2* in Song-Learning Birds and Vocal-Learning Mammals," *Journal of Heredity* 96, no. 3 (May/June 2005): 212–16, https://doi.org/10.1093/jhered/esi025; Rina Shah et al., "Expression of *FoxP2* during Zebrafish Development and in the Adult Brain," *International Journal of Developmental Biology* 50, no. 4 (March 2006): 435–38, https://doi.org/10.1387/ijdb.052065rs.

songbirds' neural circuits: Andreas R. Pfenning et al., "Convergent Transcriptional Specializations in the Brains of Humans and Song-Learning Birds," *Science* 346, no. 6215 (December 2014), https://doi.org/10.1126/science.1256846.

"a single organism as distinguished": *Merriam-Webster Dictionary*, s.v. "individual," accessed December 30, 2024, https://www.merriam-webster.com /dictionary/individual.

"the totality of an individual's behavioral": *Merriam-Webster Dictionary*, s.v. "personality," accessed December 30, 2024, https://www.merriam-webster.com /dictionary/personality.

"Parental affection, or some feeling": Darwin, *Descent of Man*, 106.

Isaac Planas-Sitjà quotes: Isaac Planas-Sitjà, in conversation with the author, September 2023.

"inherent chaos of normal development": Gerrit Arne Linneweber et al., "A Neurodevelopmental Origin of Behavioral Individuality in the *Drosophila* Visual System," *Science* 367, no. 6482 (March 6, 2020): 1112–19, https://doi.org/10 .1126/science.aaw7182.

"flies can communicate": Balint Z. Kacsoh, Julianna Bozler, and Giovanni Bosco, "Drosophila Species Learn Dialects Through Communal Living," *PLOS Genetics* 14, no 7 (July 2018): e1007430, https://doi.org/10.1371/journal.pgen .1007430.

Scott Waddell quotes: Scott Waddell, in conversation with the author, September 2023.

"Consciousness: the state of being awake": "consciousness," Dictionary, Google, accessed December 30, 2024, https://www.google.com/search?q =definition+consciousness.

"An organism has conscious mental states": Thomas Nagel, "What Is It Like to Be a Bat?," *Philosophical Review* 83, no. 4 (October 1974): 436, https://www .jstor.org/stable/2183914.

2012 Cambridge Declaration on Consciousness: Philip Low, "Cambridge Declaration on Consciousness," Proceedings of the Francis Crick Memorial Conference, Churchill College, Cambridge University, July 7, 2012, 1, https:// fcmconference.org/img/CambridgeDeclarationOnConsciousness.pdf.

2022 British Animal Welfare (Sentience) Act: Animal Welfare (Sentience) Act 2022, c. 22, https://www.legislation.gov.uk/ukpga/2022/22/enacted.

psychologist David Premack: David Premack and Guy Woodruff, "Does the Chimpanzee Have a Theory of Mind?," *Behavioral and Brain Sciences* 1, no. 4 (December 1978): 515–26, http://dx.doi.org/10.1017/S0140525X00076512.

Albany psychologist Gordon Gallup: Gordon G. Gallup Jr., "Chimpanzees: Self-Recognition," *Science* 167, no. 3914 (January 2, 1970): 86–87, http://dx .doi.org/10.1126/science.167.3914.86.

"based on rigorous, reproducible experimental evidence": Gordon G. Gallup Jr. and J. R. Anderson, "Self-Recognition in Animals: Where Do We Stand 50 Years Later? Lessons from Cleaner Wrasse and Other Species," *Psychology of Consciousness: Theory, Research, and Practice* 7, no. 1 (2020): 46–58, https:// psycnet.apa.org/doi/10.1037/cns0000206—.

the Caputo effect: Giovanni B. Caputo, "Strange-Face-In-The-Mirror Illusion," *Perception* 39, no. 7 (2010): 1007–8, https://doi.org/10.1068/p6466.

pigs are "highly sensitive animals": Donald M. Broom, Hilana Sena, and Kiera L. Moynihan, "Pigs Learn What a Mirror Image Represents and Use It to Obtain Information," *Animal Behaviour* 78, no. 5 (November 2009): 1037–41, https://doi.org/10.1016/j.anbehav.2009.07.027.

astronomer Carl Sagan: Carl Sagan, *The Dragons of Eden: Speculations on the Evolution of Human Intelligence* (Random House, 1977), 7.

many small children in non-Western countries: Tanya Broesh et al., "Cultural Variations in Children's Mirror Self-Recognition," *Journal of Cross-Cultural Psychology* 42, no. 6 (2011): 1018–29, https://doi.org/10.1177/0022022110381114.

a Scottish fold kitten named Mimo: "Curious Cat Discovers She Has Ears While Striking Pose in Mirror," Storyful Viral, September 27, 2018, YouTube video, 0:47, https://www.youtube.com/watch?v=akE2Sgg8hI8.

Diana Reiss's team marked bottlenose dolphins: Diana Reiss and Lori Marino, "Mirror Self-Recognition in the Bottlenose Dolphin: A Case of Cognitive Convergence," *PNAS* 98, no. 10 (May 2001): 5937–42, https://doi.org/10.1073/pnas.101086398.

olfactory mirror test for dogs: Alexandra Horowitz, "Smelling Themselves: Dogs Investigate Their Own Odours Longer When Modified in an 'Olfactory Mirror' Test," *Behavioural Processes* 143 (2017): 17–24, https://doi.org/10.1016/j.beproc.2017.08.001.

voles demonstrate empathy: J. P. Burkett et al., "Oxytocin-Dependent Consolation Behavior in Rodents," *Science* 351, no. 6271 (January 2016): 375–78, https://doi.org/10.1126/science.aac4785.

Likewise jays and crows: Ljerca Ostojić et al., "Evidence Suggesting That Desire-State Attribution May Govern Food-Sharing in Eurasian Jays," *PNAS* 110, no. 10 (February 2013): 4123–28, https://doi.org/10.1073/pnas.1209926110; Katy Sewall, "The Girl Who Gets Gifts from Birds," *BBC News*, February 25, 2015, https://www.bbc.com/news/magazine-31604026.

"Some of the resistance to the idea": James L. Gould, "Honey Bee Recruitment: The Dance-Language Controversy," *Science* 189, no. 4204 (August 1975): 685–93, https://doi.org/10.1126/science.1154023.

Donald Griffin persuaded naysaying scientists: Carol Kaesuk Yoon, "Donald R. Griffin, 88, Dies; Argued Animals Can Think," *New York Times*, November 14, 2003, https://www.nytimes.com/2003/11/14/nyregion/donald-r-griffin-88 -dies-argued-animals-can-think.html.

Moths and butterflies retain memories: Douglas J. Blackiston, Elena Silva Casey, and Martha R. Weiss, "Retention of Memory through Metamorphosis: Can a Moth Remember What It Learned as a Caterpillar?," *PLOS One* 3, no. 3 (March 5, 2008): e1736, https://doi.org/10.1371/journal.pone.0001736.

only way to know what it's like to be a bat: Nagel, "What Is It Like."

"If a lion could talk": Ludwig Wittgenstein, *Philosophical Investigations* (Basil Blackwell, 1958), 223.

German biologist Jacob von Uexküll: Jakob von Uexküll, *A Foray into the Worlds of Animals and Humans: With A Theory of Meaning*, trans. Joseph D. O'Neil (University of Minnesota Press, 2010), 35–36.

"Little fly": William Blake, *Songs of Innocence and of Experience* (William Blake, 1794).

CHAPTER 2: FOWL LANGUAGE

"So they show their relations to me": Walt Whitman, *Song of Myself* (Dover, 2012), 28.

"Go outside three times a day": Bernie Krause, in conversation with the author, August 2023.

Rachel Mundy quotes: Rachel Mundy, in conversation with the author, September 2023.

"What have we to do": Maurice Maeterlinck, *The Life of the Bee* (Dover, 2006), 66–67.

domestication of animals: P. A. Nikolskiy et al., "Predomestication and Wolf-Human Relationships in the Arctic Siberia of 30,000 Years Ago: Evidence from the Yana Paleolithic Site," *Stratum plus* 2018, no. 1 (2018): 231–62, https:// www.e-anthropology.com/English/Catalog/Archaeology/STM_DWL_x2MF _YjQvYfY1EVqn.aspx; "New Light Shed on the Domestication History of Sheep and Goats," Cardiff University, March 20, 2018, https://www.cardiff.ac.uk/news /view/1128739-new-light-shed-on-the-domestication-history-of-sheep-and -goats; "DNA Traces Cattle Back to a Small Herd Domesticated around 10,500 Years Ago," University College London, March 27, 2012, https://www.ucl.ac .uk/news/2012/mar/dna-traces-cattle-back-small-herd-domesticated-around -10500-years-ago; Amke Caliebe et al., "Insights into Early Pig Domestication Provided by Ancient DNA Analysis," *Scientific Reports* 7 (March 16, 2017): 44550, https://doi.org/10.1038/srep44550; Pablo Librado et al., "The Origins and Spread of Domestic Horses from the Western Eurasian Steppes," *Nature* 598 (October 2021): 634–40, https://www.nature.com/articles/s41586-021-04018 -9; Carlos A. Driscoll et al., "The Evolution of House Cats," *Scientific American*, June 2009, https://www.scientificamerican.com/article/the-taming-of-the-cat/.

evening primroses responded to bees: Marine Veits et al., "Flowers Respond to Pollinator Sound Within Minutes by Increasing Nectar Sugar Concentration," *Ecology Letters* 22, no. 9 (September 2019): 1483–92, https://doi.org/10.1111 /ele.13331.

coral larvae can be enticed: Ashlee Lillis et al., "Soundscapes Influence the Settlement of the Common Caribbean Coral *Porites astriodes* Irrespective of Light Conditions," *Royal Society Open Science* 5, no. 12 (December 2018): 181358, https://doi.org/10.1098/rsos.181358.

"the watch-dog's honest bark": Lord Byron, *Don Juan* (Henry Hoff, 1838), 52.

"the number of life goals": Sigal Zilcha-Mano, Mario Mikulincer, and Phillip R. Shaver, "Pets as Safe Havens and Secure Bases: The Moderating Role of Pet Attachment Orientations," *Journal of Research in Personality* 46, no. 5 (October 2012): 571–80, https://doi.org/10.1016/j.jrp.2012.06.005.

petting a dog can cause a transient decrease: Stanislaw Surma, Suzanne Oparil, and Krzysztof Narkiewicz, "Pet Ownership and the Risk of Arterial Hypertension and Cardiovascular Disease," *Current Hypertension Reports* 24, no. 8 (April 2022): 295–302, https://doi.org/10.1007/s11906-022-01191-8.

elderly people exposed to fish tanks: "Animals Help the Alzheimer's Disease Patient," Center for the Human-Animal Bond, Purdue University, accessed

December 15, 2024, https://vet.purdue.edu/chab/research/animals-help-the
-alzheimers-disease-patient.php.

PTSD-afflicted army veterans: Melissa Sherman et al., "Effectiveness of
Operation K9 Assistance Dogs on Suicidality in Australian Veterans with
PTSD: A 12-Month Mixed-Methods Follow-Up Study," *International Journal
of Environmental Research and Public Health* 20, no. 4 (2003): 3607, https://doi
.org/10.3390/ijerph20043607.

pet owners visit the doctor less frequently: "Pet Owners Visit Physicians Less,
Says New Report," *dvm360*, December 22, 2015, https://www.dvm360.com
/view/pet-owners-visit-physicians-less-says-new-report.

time spent with a dog: Emily R. L. Thelwell, "Paws for Thought: A Controlled
Study Investigating the Benefits of Interacting with a House-Trained Dog on
University Students Mood and Anxiety," *Animals* 9, no. 10 (October 2019): 846,
https://doi.org/10.3390/ani9100846.

a 2001 study of "hypertensive stockbrokers": Lois Baker, "Pet Dog or Cat
Controls Blood Pressure Better Than ACE Inhibitor, UB Study of Stockbrokers
Finds," News Center, University at Buffalo, November 7, 1999, https://www
.buffalo.edu/news/releases/1999/11/4489.html.

among pet-owning couples: Karen Allen, Jim Blascovich, and Wendy B.
Mendes, "Cardiovascular Reactivity and the Presence of Pets, Friends, and
Spouses: The Truth about Cats and Dogs," *Psychosomatic Medicine* 64, no.
5 (September-October 2002): 727–39, https://doi.org/10.1097/01.psy
.0000024236.11538.41.

twice as many pet owners died: Hal Herzog, "Will Getting a Pet Make You
Healthier?," *Psychology Today*, August 14, 2011, https://www.psychologytoday
.com/ie/blog/animals-and-us/201108/will-getting-pet-make-you-healthier.

"Secure attachment, strong emotional bonds": Karoline Gerwisch et al.,
"A Pilot Study into the Effects of PTSD-Assistance Dogs' Work on Their
Salivary Cortisol Levels and Their Handlers' Quality of Life," *Journal of Applied
Animal Welfare Science* (2023): 1–13, https://doi.org/10.1080/10888705.2023
.2259795.

Dogs produce oxytocin: Sanni Somppi et al., "Nasal Oxytocin Treatment Biases Dogs' Visual Attention and Emotional Response toward Positive Human Facial Expressions," *Frontiers in Psychology* 8 (October 2017): 1854, https://doi.org/10.3389/fpsyg.2017.01854; Miho Nagasawa et al., "Oxytocin-Gaze Positive Loop and the Coevolution of Dog-Human Bonds," *Science* 348, no. 6232 (April 2015): 333–36, https://doi.org/10.1126/science.1261022.

listening to recorded birdsong: E. Stobbe et al., "Birdsongs Alleviate Anxiety and Paranoia in Healthy Participants," *Scientific Reports* 12 (2022): 16414, https://doi.org/10.1038/s41598-022-20841-0——.

a 2022 report from an EU-sponsored project: Elodie F. Briefer et al., "Classification of Pig Calls Produced from Birth to Slaughter According to Their Emotional Valence and Context of Production," *Scientific Reports* 12 (2022): 3409, https://doi.org/10.1038/s41598-022-07174-8——.

"Knowing what these vocalizations mean": "Media Opportunity: Fowl Language: Dalhousie University Researcher Uses AI to Crack the Code of Clucks and Unravel Chicken Chatter Secrets, Opening the Door to an Improved Quality of Life," Media Centre, Dalhousie University, March 25, 2024, https://www.dal.ca/news/media/media-releases/2024/03/25/media_opportunity__fowl_language__dalhousie_university_researcher_uses_ai_to_crack_the_code_of_clucks_and_unravel_chicken_chatter_secrets__opening_the_door_to_an_improved_quality_of_life.html.

a 2024 Queensland University study: Nicky McGrath et al., "Humans Can Identify Reward-Related Call Types of Chickens," *Royal Society Open Science* 11, no. 1 (January 2024): 231284, https://doi.org/10.1098/rsos.231284.

Peter Singer quotes: Peter Singer, in conversation with the author, October 2023.

Marc Bekoff quotes: Marc Bekoff, in conversation with the author, October 2023.

Sometimes called a slow blink: "Cat Slow Blinking: Why Your Cat Blinks Slowly at You," Veterinary Healthcare Associates, accessed December 16, 2024, https://vhavets.com/blog/cat-slow-blinking/.

interactions between zookeepers and their charges: Geoff Hosey and Vicky Melfi, "Human-Animal Bonds Between Zoo Professionals and the Animals in Their Care," *Zoo Biology* 31, no. 1 (January/February 2012): 13–26, https://doi.org/10.1002/zoo.20359; Kate C. Baker, "Benefits of Positive Human Interaction for Socially Housed Chimpanzees," *Animal Welfare* 13, no. 2 (May 2004): 239–45, https://pmc.ncbi.nlm.nih.gov/articles/PMC2875797/.

how canine behavioral euthanasia could be prevented: Miranda Hitchcock et al., "Factors Associated with Behavioral Euthanasia in Pet Dogs," *Frontiers in Veterinary Science* 11 (April 2024): 1387076, https://doi.org/10.3389/fvets.2024.1387076.

Jonathan Balcombe quotes: Jonathan Balcombe, in conversation with the author, October 2023.

mice serenade prospective partners: Joshua Neunuebel et al., "Female Mice Ultrasonically Interact with Males during Courtship Displays," *eLife* 4 (2015): e06203, https://doi.org/10.7554/eLife.06203.

they surgically deafened mice: Gustavo Arriaga, Eric P. Zhou, and Erich D. Jervis, "Of Mice, Birds, and Men: The Mouse Ultrasonic Song System Has Some Features Similar to Humans and Song-Learning Birds," *PLOS One* 7, no. 10 (October 2012): e46610, https://doi.org/10.1371/journal.pone.0046610.

Sounding Soil project: Sounding Soil, accessed February 18, 2025, https://www.soundingsoil.ch/en/.

cicadas and katydids: Chunpeng Xu et al., "High Acoustic Diversity and Behavioral Complexity of Katydids in the Mesozoic Soundscape," *PNAS* 119, no. 51 (December 2022): e2210601119, https://doi.org/10.1073/pnas.2210601119.

behavioral ecologist Rachel Grant: Rachel A. Grant, Jean Pierre Raulin, and Friedmann T. Freund, "Changes in Animal Activity Prior to a Major ($M=7$) Earthquake in the Peruvian Andes," *Physics and Chemistry of the Earth, Parts A/B/C* 85–86 (2015): 69–77, https://doi.org/10.1016/j.pce.2015.02.012.

French project that sought to track migratory birds: Kivi Kuaka, accessed February 18, 2025, https://kivikuaka.fr/?lang=en%2F.

in earthquake-prone central Italy: Martin Wikelski et al., "Potential Short-Term Earthquake Forecasting by Farm Animal Monitoring," *Ethology* 126, no. 9 (September 2020): 931–41, https://doi.org/10.1111/eth.13078.

California State Earthquake Investigation Commission: "The California Earthquake of April 18, 1906: Report of the State Earthquake Investigation Commission, in two volumes and atlas," 382–83, Online Archive of California, https://oac.cdlib.org/view?docId=hb1h4n989f;NAAN=13030&doc.view =frames&chunk.id=div00123&toc.id=div00006&brand=eqf\.

"For five days before Helice disappeared": Aelian, *On the Characteristics of Animals*, books VI–XI (William Heinemann, 1959), 387.

a 2005 edition of *National Geographic*: Maryann Mott, "Did Animals Sense Tsunami Was Coming?," *National Geographic*, January 4, 2005, https://www .nationalgeographic.com/animals/article/news-animals-tsunami-sense-coming?.

United States Geological Survey: "Can Animals Predict an Earthquake?," United States Geological Survey, accessed December 17, 2024, https://www .usgs.gov/programs/earthquake-hazards/animals-earthquake-prediction; "Earthquake Facts & Earthquake Fantasy," United States Geological Survey, accessed December 17, 2024, https://www.usgs.gov/programs/earthquake -hazards/earthquake-facts-earthquake-fantasy.

CHAPTER 3: A ROUND OF PIGTIONARY

"We have two ears and one mouth": Diogenes Laërtius, *The Lives and Opinions of Eminent Philosophers* (George Bell & Sons, 1895), 268.

"Let there be no exhibition of anger": Cicero, *De Officiis* (William Heinemann, 1908), 139–40.

Dogs are excellent listeners: Laura V. Cuaya et al., "Speech Naturalness Detection and Language Representation in the Dog Brain," *NeuroImage* 248 (March 2022): 118811, https://doi.org/10.1016/j.neuroimage.2021.118811.

"increased activity in primary reward regions": A. Andics et al., "Neural Mechanisms for Lexical Processing in Dogs," *Science* 353, no. 6303 (August 2016): 1030–32, https://doi.org/10.1126/science.aaf3777.

Dogs are exceptional at reading our body language: Natalia Albuquerque et al., "Dogs Recognize Dog and Human Emotions," *Biology Letters* 12, no. 1 (January 2016): 20150883, https://doi.org/10.1098/rsbl.2015.0883; Miho Nagasawa et al., "Dogs Can Discriminate Human Smiling Faces from Blank Expressions," *Animal Cognition* 14, no. 4 (July 2011): 525–33, https://doi.org /10.1007/s10071-011-0386-5; Juliane Kaminski et al., "Human Attention Affects Facial Expressions in Domestic Dogs," *Scientific Reports* 7, no. 1 (October 2017): 12914, https://doi.org/10.1038/s41598-017-12781-x.

dogs' barks originally developed: Alexandra Horowitz, "Why Puppies Bark," *Time*, September 23, 2022, https://time.com/6215910/why-puppies-bark/.

Wolves rarely bark: "The Language of Wolves," Living With Wolves Museum, accessed December 20, 2024, https://www.livingwithwolves.org/about-wolves /language/.

humans can automatically understand barks: Péter Pongrácz et al., "Human Listeners Are Able to Classify Dog (*Canis familiaris*) Barks Recorded in Different Situations," *Journal of Comparative Psychology* 119, no. 2 (2005): 136–44, https://psycnet.apa.org/doi/10.1037/0735-7036.119.2.136.

Alexandra Horowitz quotes: Alexandra Horowitz, in conversation with the author, November 2023.

astonishing sense of smell: Howard Lancum, *Badgers' Year* (C. Lockwood & Son, 1954), 28.

Dogs can smell emotions: Biaggio D'Aniello et al., "Interspecies Transmission of Emotional Information via Chemosignals: From Humans to Dogs (*Canis lupus familiaris*)," *Animal Cognition* 21 (October 2017): 67–78, https://doi .org/10.1007/s10071-017-1139-x—; Paula Jendrny et al., "Canine Olfactory Detection and Its Relevance to Medical Detection," *BMC Infectious Diseases* 21 (2021): 838, https://doi.org/10.1186/s12879-021-06523-8; Clara Wilson et al., "Dogs Can Discriminate between Human Baseline and Psychological Stress Condition Odours," *PLOS One* 17, no. 9 (2022): e0274143, https://doi.org/10 .1371/journal.pone.0274143.

Songbirds dream in songs: Juan F. Döppler et al., "Synthesizing Avian Dreams,"

Chaos, An Interdisciplinary Journal of Nonlinear Science 34, no. 4 (April 2024): 043103, https://doi.org/10.1063/5.0194301.

indigenous and riverine communities: Chanelle Dupuis, "The Smell of Water: A Liquid Witness to Environmental Change in the Amazon," *Gastronomica* 22, no. 4 (November 2022): 1–9, https://doi.org/10.1525/gfc.2022.22.4.1.

"dogs may be better than humans": John P. McGann, "Poor Human Olfaction Is a Nineteenth Century Myth," *Science* 356, no. 6338 (May 2017), https://doi .org/10.1126/science.aam7263.

an experiment wherein thirty-two Californian students: Jess Porter et al., "Mechanisms of Scent-Tracking in Humans," *Nature* 10, no. 1 (February 2007): 27–29, https://doi.org/10.1038/nn1819.

Irene Pepperberg quotes: Irene Pepperberg, in conversation with the author, October 2023.

Pamela Clark quotes: Pamela Clark, in conversation with the author, October 2023.

"oldest Guinness-World-Record-certified parrot": "Oldest parrot ever,"Guinness World Records, accessed March 10, 2025, https://www.guinnessworldrecords .com/world-records/442525-oldest-parrot-ever#:~:text=The%20oldest%20parrot %20ever%20is,Brookfield%20Zoo%20in%20May%201934.

Con Slobodchikoff quotes: Con Slobodchikoff, in conversation with the author, October 2023.

"Thus we ultimately reach the conclusion": Jakob von Uexküll, "A Stroll Through the Worlds of Animals and Men: A Picture Book of Invisible Worlds," in *Instinctive Behavior: The Development of a Modern Concept*, trans. and ed. Claire H. Schiller (International Universities Press, 1957), 5–80.

elephant ethogram: "The Elephant Ethogram," Elephant Voices, accessed December 21, 2024, https://www.elephantvoices.org/elephant-ethogram.html.

"high-duration tail movement": Miriam Marcet Rius et al., "Tail and Ear Movements as Possible Indicators of Emotions in Pigs," *Applied Animal Behaviour Science* 205 (August 2018): 14–18, https://doi.org/10.1016/j.applanim.2018.05 .012.

motivational structure hypothesis: Eugene S. Morton, "On the Occurrence and Significance of Motivation-Structural Rules in Some Bird and Mammal Sounds," *American Naturalist* 111, no. 981 (September-October 1977): 855–69, https://doi.org/10.1086/283219.

four million pigs are slaughtered: Alex Thornton, "This Is How Many Animals We Eat Each Year," World Economic Forum, February 8, 2019, https://www .weforum.org/stories/2019/02/chart-of-the-day-this-is-how-many-animals-we -eat-each-year/.

CHAPTER 4: STILL WATERS

"To me, nothing is voiceless": *Simona*, directed by Natalia Koryncka-Gruz (Eureka Studio, 2022), https://dafilms.com/film/15924-simona.

"Silence is of different kinds": Charlotte Brontë, *Villette* (Smith, Elder, 1857), 334.

archerfish can be trained: Cait Newport et al., "Discrimination of Human Faces by Archerfish (*Toxotes chatareus*)," *Scientific Reports* 6 (2016): 27523, https://doi .org/10.1038/srep27523.

"The octopus is a stupid creature": Aristotle, *Historia Animalium* 9.37.622, https://archive.org/details/historiaanimaliu00aris_0/page/n443/mode/2up? =622.

"What sort of space": Henry David Thoreau, *Walden* (Walter Scott, 1888), 131.

Ahmad Abdella quotes: Dr. Ahmad Abdella, in conversation with the author, November 2023.

dwarf cuttlefish can pass the Stanford marshmallow test: Alexandra K. Schnell et al., "Cuttlefish Exert Self-Control in a Delay of Gratification Test," *Proceedings of the Royal Society B* 288, no. 1946 (March 2021): 20203161, https://doi.org /10.1098/rspb.2020.3161.

giant Pacific octopuses can recognize human faces: Roland C. Anderson et al., "Octopuses (*Enteroctopus dofleini*) Recognize Individual Humans," *Journal of Applied Animal Welfare Science* 13, no. 3 (June 2010): 261–72, https://doi.org /10.1080/10888705.2010.483892.

Alex Schnell quotes: Dr. Alex Schnell, in conversation with the author, November 2023.

grouper and coral trout: Alexander L. Vail, Andrea Manica, and Redouan Bshary, "Referential Gestures in Fish Collaborative Hunting," *Nature Communications* 4 (April 2013): 1765, https://doi.org/10.1038/ncomms2781.

octopuses punching interspecies hunting partners: Eduardo Sampaio et al., "Multidimensional Social Influence Drives Leadership and Composition-Dependent Success in Octopus-Fish Hunting Groups," *Nature Ecology & Evolution* 8 (September 2024): 2072–84, https://doi.org/10.1038/s41559-024 -02525-2——.

octopuses use tools: Julian K. Finn, Tom Tregenza, and Mark D. Norman, "Defensive Tool Use in a Coconut-Carrying Octopus," *Current Biology* 19, no. 23 (December 2009): 1069–70, https://doi.org/10.1016/j.cub.2009.10.052.

"It is just like man's vanity": Mark Twain, *What Is Man? And Other Essays* (Harper & Brothers, 1917), 101.

whalesongs change: Ellen C. Garland and Peter K. McGregor, "Cultural Transmission, Evolution and Revolution in Vocal Displays: Insights from Bird and Whale Song," *Frontiers in Psychology* 11 (2020): 544929, https://doi.org/10 .3389/fpsyg.2020.544929.

a paper titled "Interactive Bioacoustic Playback": Brenda McCowan et al., "Interactive Bioacoustic Feedback as a Tool for Detecting and Exploring Nonhuman Intelligence: 'Conversing' with an Alaskan Humpback Whale," *PeerJ* 11 (November 2023): e16349, https://doi.org/10.7717/peerj.16349.

"An important assumption": "Whale-SETI: Groundbreaking Encounter with Humpback Whales Reveals Potential for Non-Human Intelligence Communication," SETI Institute, December 12, 2023, https://www.seti.org /press-release/whale-seti-groundbreaking-encounter-humpback-whales-reveals -potential-non-human-intelligence.

scientists broadcast playback: Michelle E. H. Fournet, Andy Szabo, and David Mellinger, "Repertoire and Classification of Non-Song Calls in Southeast Alaskan Humpback Whales (*Megaptera novaeangliae*)," *Journal of Acoustical*

Society of America 137, no. 1 (January 2015): 1–10, http://dx.doi.org/10.1121/1.4904504.

"Scientists Share World's First 'Conversation'": Nikki Main, "Scientists Share World's First 'Conversation' Between Humans and Whales—and Say It's the First Step to Understanding Aliens," *Daily Mail,* April 17, 2024, https://www.dailymail.co.uk/sciencetech/article-13319043/worlds-conversation-humans-whales-aliens.html.

Roger Payne published an impassioned article: Roger Payne, "I Spent My Life Saving the Whales. Now They Might Save Us," *Time,* June 5, 2023, https://time.com/6284884/whale-scientist-last-please-save-the-species/.

Denise Herzing quotes: Dr. Denise Herzing, in conversation with the author, December 2023.

zoologist Katy Payne: "Katy Payne On Elephants," Elephant Listening Project, Cornell University, accessed January 3, 2025, https://www.elephantlisteningproject.org/katy-payne-on-elephants/.

these conversations can travel: Nicholas St. Fleur, "An Elephant's Silent Call," *Science,* August 2, 2012, https://www.science.org/content/article/elephants-silent-call.

biologist Camila Ferrara discovered: Gabriel Jorgewich-Cohen et al., "Prehatch Sounds and Coordinated Birth in Turtles," *Ecology and Evolution* 14, no. 10 (October 2024): e70410, https://doi.org/10.1002/ece3.70410.

happy dogs tend to wag their tails: Marcelo Siniscalchi et al., "Seeing Left- or Right-Asymmetric Tail Wagging Produces Different Emotional Responses in Dogs," *Current Biology* 23, no. 22 (November 2013): 2279–82, https://doi.org/10.1016/j.cub.2013.09.027.

Meowtalk: Meowtalk, accessed February 19, 2025, https://www.meowtalk.app.

domestic felines can distinguish: Charlotte de Mouzon, Marine Gonthier, and Gérard Leboucher, "Discrimination of Cat-Directed Speech from Human-Directed Speech in a Population of Indoor Companion Cats (*Felis catus*)," *Animal Cognition* 26 (2023): 611–19, https://doi.org/10.1007/s10071-022-01674-w.

cats' voices are more often directed: Grace Carroll, "Cats Love to Meow at Humans. Now We Know Why," *Live Science*, July 7, 2024, https://www .livescience.com/animals/cats-love-to-meow-at-humans-now-we-know-why.

one 2020 study suggests we owners: Emanuela Prato-Previde et al., "What's in a Meow? A Study on Human Classification and Interpretation of Domestic Cat Vocalizations," *Animals* 2, no. 12 (December 2020): 2390, https://doi.org/10 .3390/ani10122390.

Meowsic: Melody in Human-Cat Communication (Meowsic), accessed February 19, 2025, https://meowsic.se.

"This basic scientific method": "Ethology and Stress Diseases," Nobel Lecture, December 12, 1973, https://www.nobelprize.org/uploads/2018/06/tinbergen -lecture.pdf.

"And this our life": Shakespeare, William. *As You Like It*. In *The Complete Works of William Shakespeare*, edited by W. G. Clark and W. Aldis Wright (Holt, Rinehart and Winston, 1952), 786–799.

lawyer who used a ChatGPT-created judgment: Molly Bohannon, "Lawyer Used ChatGPT In Court—And Cited Fake Cases. A Judge is Considering Sanctions," *Forbes*, June 8, 2023, https://www.forbes.com/sites/mollybohannon /2023/06/08/lawyer-used-chatgpt-in-court-and-cited-fake-cases-a-judge-is -considering-sanctions/.

Temple Grandin quotes: Dr. Temple Grandin, in conversation with the author, December 2023.

CHAPTER 5: TREADING THE BOARDS AND LESSONS FROM APES

"For it is the destiny": John Stewart Collis, *The Worm Forgives the Plough* (Vintage, 2009), 180.

apes involved in language research: "Science: Are Those Apes Really Talking?," *Time*, March 10, 1980, https://time.com/archive/6883428/science-are-those -apes-really-talking/; "Sesame Street: Nim Chimpsky Matches Teddy Bear's Nose (1978)," Treble Gnocchi, July 24, 2024, YouTube video, 1:19, https:// www.youtube.com/watch?v=C7NByacPbVA.

apes' vocal mechanisms: Claude Desrochers, "Communication Between Chimpanzee Residents and Human Caregivers," Fauna Foundation, April 25, 2022, https://faunafoundation.org/news-events/blog-pages/chimps /communication-between-chimpanzee-residents-and-human-caregivers .html.

U.S. National Institutes of Health discontinued medical testing: "NIH Will No Longer Support Biomedical Research on Chimpanzees," National Institutes of Health, U.S. Department of Health and Human Services, November 17, 2015, https://www.nih.gov/about-nih/who-we-are/nih-director/statements/nih -will-no-longer-support-biomedical-research-chimpanzees.

everyday people are excellent at discerning: Kirsty E. Graham and Catherine Hobaiter, "Towards a Great Ape Dictionary: Inexperienced Humans Understand Common Nonhuman Ape Gestures," *PLOS Biology* 21, no. 1 (January 2023): e3001939, https://doi.org/10.1371/journal.pbio.3001939.

Cat Hobaiter quotes: Cat Hobaiter, in conversation with the author, January 2024.

Elizabeth Hess's 2008 book: Elizabeth Hess, *Nim Chimpsky: The Chimp Who Would be Human* (Bantam, 2008), 290.

Herb Terrace quotes: Herb Terrace, in conversation with the author, January 2024.

horses can be taught to touch: Cecilie M. Mejdell et al., "Horses Can Learn to Use Symbols to Communicate Their Preferences," *Applied Animal Behaviour Science* 184 (November 2016): 66–73, https://doi.org/10.1016/j.applanim .2016.07.014.

Bunny on Instagram and TikTok: Alexis Devine (@whataboutbunny), "Well this is certainly the existential content that TikTok had been hoping for," Instagram, November 14, 2020, https://www.instagram.com/p /CHk6GzyhReG/; Alexis Devine (@whataboutbunny), "Check your doggy's paws if you live in an area with foxtail," TikTok, September 7, 2020, https:// www.tiktok.com/@whataboutbunny/video/6880896395230874886?lang=en; Fluentpet (@fluentpet), "A timeless Bunny classic—'Belly Go Bird Belly'—to kick off the #Thanksgiving Festivities!," Instagram, November 23, 2022, https:// www.instagram.com/p/ClTpLAfsL6g/.

Leo Trottier quotes: Leo Trottier, in conversation with the author, January 2024.

A recent study of button boards: Justyna Wlodarczyk et al., "Talking Dogs: The Paradoxes Inherent in the Cultural Phenomenon of Soundboard Use by Dogs," *Animals* 14, no. 22 (2024): 3272, https://doi.org/10.3390 /ani14223272.

fire ants combined pheromones: Richard Rhodes, "Speaking Pheromone," *Harvard Magazine*, November-December 2021, https://www.harvardmagazine .com/2021/10/feature-speaking-pheomone.

researchers successfully trained a captive orca: José Z. Abramson et al., "Imitation of Novel Conspecific and Human Speech Sounds in the Killer Whale (*Orcinus orca*)," *Proceedings of the Royal Society B* 285, no. 1871 (January 2018): 20172171, https://doi.org/10.1098/rspb.2017.2171.

"John Burroughs, who's a shark on birds": James J. Montague, *More Truth Than Poetry* (George H. Doran, 1920), 110–11, https://archive.org/details /moretruththanpo00montgoog/page/n113/mode/2up.

"Negative Capability": John Keats, *Selected Letters* (Oxford University Press, 2002), 40.

CHAPTER 6: MAYBE SHE'S THIRSTY

a 2017 Durham University study: Caterina Spiezio et al., "Does Positive Reinforcement Training Affect the Behavior and Welfare of Zoo Animals? The Case of the Ring-Tailed Lemur (*Lemur catta*)," *Applied Animal Behaviour Science* 196 (November 2017): 91–99, https://doi.org/10.1016/j.applanim.2017.07 .007.

"Everything that irritates": Carl G. Jung, *Memories, Dreams, Reflections* (Vintage, 1989), 247.

"I'll have a starling": William Shakespeare, *King Henry IV Parts One and Two* (President, 1909), 32.

"One day ye Bishop of Canterbury": "Department of Quotation: The

Talking Starling," *Two Nerdy History Girls* (blog), October 25, 2009, https:// twonerdyhistorygirls.blogspot.com/2009/10/department-of-quotation-talking .html.

"What a horse does under compulsion": Xenophon, *The Art of Horsemanship*, trans. Morris H. Morgan (Little, Brown, 1893), 62.

"keep up their spirits": Xenophon, *The Sportsman: On Hunting, A Sportsman's Manual, Commonly Called Cynegeticus*, trans. Henry Dakyns (Project Gutenberg, 1998), chapter 6, https://www.gutenberg.org/cache/epub/1180/pg1180-images .html.

"Never forget to have some delicacy": W. N. Hutchinson, *Dog Breaking* (John Murray, 1865), 203.

cows enjoy listening to classical music: L.-M. Erasmus et al., "Exploring the Effect of Auditory Stimuli on Activity Levels, Milk Yield and Faecal Glucocortinoid Metabolite Concentrations in Holstein Cows," *Domestic Animal Endocrinology* 82 (January 2023): 106767, https://doi.org/10.1016/j .domaniend.2022.106767; K. Uetake, J. F. Hurnik, and L. Johnson, "Effect of Music on Voluntary Approach of Dairy Cows to an Automatic Milking System," *Applied Animal Behaviour Science* 53, no. 3 (June 1997):175–82, https://doi.org /10.1016/S0168–1591(96)01159–8.

A quick YouTube search: "Sake, the sea lion, on The Tonight Show with Dr. Jenifer Zeligs," Animal Training and Research International, July 16, 2017, YouTube video, 7:11, https://www.youtube.com/watch?v=Z6jdhBUPPeA.

Hans invariably provided the correct answer: "Berlin's Wonderful Horse: He Can Do Almost Everything but Talk—How He Was Taught," *New York Times*, September 4, 1904, https://timesmachine.nytimes.com/timesmachine/1904/09 /04/101396572.pdf.

Jenifer Zeligs quotes: Jenifer Zeligs, in conversation with the author, February 2024.

the Gaucho Derby: Gaucho Derby, accessed February 20, 2025, https:// equestrianists.com/gaucho-derby/.

polyvagal theory: Stephen W. Porges, "Polyvagal Theory: A Science of Safety,"

Frontiers in Integrative Neuroscience 16 (2022): 871227, https://doi.org/10.3389 /fnint.2022.871227.

Richard Erskine's definition of attunement: Richard Gordon Erskine, "Attunement and Involvement: Therapeutic Responses to Relational Needs," *International Journal of Psychotherapy* 3, no. 3 (November 1998): 235–44, https://www.proquest.com/docview/212046787.

Warwick Schiller quotes: Warwick Schiller, in conversation with the author, February 2024.

"While with an eye made quiet by the power": William Wordsworth, *The Complete Poetical Works of William Wordsworth*, II, 1798–1800 (Houghton Mifflin, 1919), 87.

"Who has seen the wind?": Christina Rossetti, "Who Has Seen the Wind?," Poetry Foundation, accessed January 5, 2025, https://www.poetryfoundation .org/poems/43197/who-has-seen-the-wind.

the methods of Dr. Bob Bailey: Tom Vanderbilt, "The CIA's Most Highly-Trained Spies Weren't Even Human," *Smithsonian Magazine,* October 2023, https://www.smithsonianmag.com/history/the-cias-most-highly-trained-spies -werent-even-human-20149/.

"because success itself is reinforcing": "Learning from Cats and Seagulls… Chat with Bob Bailey," *Susan Garrett's Dog Training Blog*, May 13, 2019, https:// susangarrettdogagility.com/2019/05/chat-with-bob-bailey/.

a just-posted Airbnb review: "Oughterson Plantation—The Barn Villa," Airbnb, accessed January 5, 2025, https://www.airbnb.ca/rooms/7430598.

CHAPTER 7: KNOW THYSELF

"Public opinion is a weak tyrant": Thoreau, *Walden*, 6.

"Two tears in a bucket": John Berendt, *Midnight in the Garden of Good and Evil* (Vintage, 1999), 104.

"The less we think": "The Hidden Magic of the Animal-Human Relationship—James French—TEDx Bologna," TEDx Talks, November 2, 2023, YouTube video, 17:42, https://www.youtube.com/watch?v=dKDAtEZpUY0.

"Half of building trust": "Discover the Mindful Approach That Helps You as Much as Your Animal," Trust Technique, accessed January 5, 2025, https://trust -technique.com/product/messages-of-trust/.

"Every being is seeking connection": French, "Hidden Magic."

The average horse goes through seven homes: Julie Castle, "The Growing Movement to Help Homeless Horses," Best Friends, May 10, 2011, https:// bestfriends.org/stories/julie-castle-blog/growing-movement-help-homeless -horses.

when riders were told an umbrella: Linda J. Keeling, Liv Jonare, and Lovisa Lanneborn, "Investigating Horse-Human Interactions: The Effect of a Nervous Human," *Veterinary Journal* 181, no. 1 (July 2009): 70–71, https://doi.org/10 .1016/j.tvjl.2009.03.013.

when people talk or share an experience: Lydia Denworth, "Brain Waves Synchronize when People Interact," *Scientific American* 329, no.1 (July/August 2023): 50, https://www.scientificamerican.com/article/brain-waves-synchronize -when-people-interact/.

a 2019 study examining emotional contagion: Maki Katayama et al., "Emotional Contagion of Humans to Dogs Is Facilitated by Duration of Ownership," *Frontiers in Psychology* 10 (July 2019): 1678, https://doi.org/10 .3389/fpsyg.2019.01678.

Californian HeartMath Institute observed similar results: "A Boy And His Dog—Heart Rhythm Entrainment," Heart Math Institute, accessed January 5, 2025, https://www.heartmath.com/inspire-a-change-of-heart/.

box breathing favored by Navy SEALs: Noma Nazish, "How to De-Stress in 5 Minutes or Less, According to a Navy SEAL," *Forbes*, May 30, 2019, https:// www.forbes.com/sites/nomanazish/2019/05/30/how-to-de-stress-in-5-minutes -or-less-according-to-a-navy-seal/.

Susan Fay quotes: Dr. Susan Fay, in conversation with the author, March 2024.

"I am not where my body is": Henry D. Thoreau, *Walking* (Riverside Press, 1914), 17.

CHAPTER 8: "HAPPY THEM UP"

"Through primrose tufts": Lynne McMahon and Averill Curdy, *The Longman Anthology of Poetry* (Pearson, 2006), 658.

"All animals, except man": Samuel Butler, *The Way of All Flesh* (E. P. Dutton, 1916), 92.

Play comes early to humans: "How Your Baby's Sense of Humour Develops— And What You Can Do to Boost It," BBC, accessed January 5, 2025, https:// www.bbc.co.uk/tiny-happy-people/articles/zjtsvk7.

goats can distinguish between happy and unhappy: Luigi Baciadonna et al., "Goats Distinguish Between Positive and Negative Emotion-Linked Vocalisations," *Frontiers in Zoology* 16, no. 25 (2019), https://doi.org/10.1186 /s12983-019-0323-z.

"sense of beauty": Darwin, *Descent of Man*, 65.

cows moo in regional accents: "Cows Also Have 'Regional Accents,'" *BBC News*, August 23, 2006, http://news.bbc.co.uk/2/hi/5277090.stm.

Ronald Rongen quotes: Ronald Rongen, in conversation with the author, April 2024.

amounts of manure: Rhonda Miller, "How Much Manure Will My Animals Produce?," Utah State University, accessed January 5, 2025, https://extension .usu.edu/smallfarms/files/How_Much_Manure.pdf.

"Of all quadrupeds": Aristotle, *Historia Animalium* 9.3.610, https://archive .org/details/historiaanimaliu00aris_0/page/n405/mode/2up?q=610.

sheep experience emotional contagion: Tomohiro Yonezawa et al., "Presence of Contagious Yawning in Sheep," *Animal Science Journal* 88, no. 1 (September 8, 2016): 195–200, https://doi.org/10.1111/asj.12681.

positive experiences make them optimistic: Alexandra Destrez et al., "Chronic Stress Induces Pessimistic-Like Judgment and Learning Deficits in Sheep," *Applied Animal Behaviour Science* 148, nos. 1–2 (September 2013): 28–36, https://doi.org/10.1016/j.applanim.2013.07.016.

sheep have strong social bonds: Laura Ozella et al., "Association Networks and Social Temporal Dynamics in Ewes and Lambs," *Applied Animal Behaviour*

Science 246 (January 2022): 105515, https://doi.org/10.1016/j.applanim.2021
.105515.

lambs love being petted: Marjorie Coulon et al., "Do Lambs Perceive Regular
Human Stroking as Pleasant? Behavior and Heart Rate Variability Analysis,"
PLOS One 10, no. 2 (February 2015): e0118617, https://doi.org/10.1371
/journal.pone.0118617.

sheep recognize human faces: Franziska Knoll, Rita P. Goncalves, and A.
Jennifer Morton, "Sheep Recognize Familiar and Unfamiliar Human Faces from
Two-Dimensional Images," *Royal Society Open Science* 4, no. 11 (November
2017): 171228, https://doi.org/10.1098/rsos.171228.

"most tricks performed by a sheep in a minute": Vicki Newman, "Animal Lover
Claims Records with Horse and Sheep Who Can Perform Hundreds of Tricks,"
Guinness World Records, August 30, 2023, https://www.guinnessworldrecords
.com/news/2023/8/animal-lover-claims-records-with-horse-and-sheep-who-can
-perform-hundreds-of-tric-757554.

"Almost all creativity": Scott Barry Kaufman (@sbkaufman), "'Almost all creativity involves purposeful play'- Abraham Maslow," X, July 10, 2020, https://x
.com/sbkaufman/status/1281740235831832576.

bumblebees play: Hiruni Samadi Galpayage Dona et al,, "Do Bumble Bees
Play?," *Animal Behaviour* 194 (December 2022): 239–51, http://dx.doi.org/10
.1016/j.anbehav.2022.08.013.

most adorable footage: "Watch bumblebees play with toys simply for fun," New
Scientist, November 2, 2022, YouTube video, 0:29, https://www.youtube.com
/watch?v=oOKK_KKiKgE.

"Males do absolutely no useful work": Lars Chittka, "The Mind of a Bee,"
Interspecies Internet, October 28, 2023, https://www.interspecies.io/lectures
/the-mind-of-a-bee.

"Lo and behold, it sounded": Mark Zaborney, "Jaak Panksepp (1943–2017);
BGSU Researcher Recognized for Work with Emotions, Brain," *Toledo Blade*,
April 20, 2017, https://www.toledoblade.com/news/deaths/2017/04/20/Jaak
-Panksepp-1943–2017-BGSU-researcher-recognized-for-work-with-emotions

-brain/stories/20170419293.

"We got totally addicted to this": Pamela Weintraub, "Discover Interview: Jaak Panksepp Pinned Down Humanity's 7 Primal Emotions," *Discover*, May 30, 2012, https://www.discovermagazine.com/mind/discover-interview-jaak -panksepp-pinned-down-humanitys-7-primal-emotions.

conservationist and scuba diver Jim Abernethy: "Tiger Shark Love," Jim Abernethy, November 29, 2014, YouTube video, 2:31, https://www.youtube .com/watch?v=gEu1qsFMOdk.

sharks have friends: Yannis P. Papastamatiou et al., "Multiyear Social Stability and Social Information Use in Reef Sharks with Diel Fission-Fusion Dynamics," *Proceedings of the Royal Society B* 287, no 1932 (August 2020): 20201063, https:// doi.org/10.1098/rspb.2020.1063.

sharks have personalities: "What Is Personality?," Bimini Shark Lab, accessed January 5, 2025, https://www.biminisharklab.com/shark-personality.

"There's no such thing as shark-infested water": "Meet Andy Nosal, the Scripps Researcher With a Passion for Science Communication—And Sharks," Scripps Institution of Oceanography, UC San Diego, January 12, 2016, https:// scripps.ucsd.edu/news/meet-andy-nosal-scripps-researcher-passion-science -communication-sharks.

"Conditioning and society": "Talking With the Ancient Ones," Animal Thoughts, accessed January 5, 2025, https://animalthoughts.com/talking-with -the-ancient-ones/.

Aimee Morgana quotes: Aimee Morgana, email to the author, April 2024.

"Dogs shew what may be fairly called": Darwin, *Descent of Man*, 49.

training her family's famous border collie: John W. Pilley, "Border Collie Comprehends Sentences Containing a Prepositional Object, Verb, and Direct Object," *Learning And Motivation* 4, no. 44 (November 2013): 229–40, https:// doi.org/10.1016/j.lmot.2013.02.003.

Pilley Bianchi quotes: Pilley Bianchi, in conversation with the author, April 2024.

a 2016 study found bumblebees experience: David Baracchi, Mathieu

Lihoreau, and Martin Guirfa, "Do Insects Have Emotions? Some Insights from Bumble Bees," *Frontiers in Behavioural Neuroscience* 11 (August 2017): 157, https://doi.org/10.3389/fnbeh.2017.00157.

"The wrinkled sea": Alfred Lord Tennyson, *The Complete Works of Alfred Lord Tennyson* (Frederick A. Stokes, 1891), 101.

horses understand our facial expressions: Miléna Trösch et al., "Horses Categorize Human Emotions Cross-Modally Based on Facial Expression and Non-Verbal Vocalizations," *Animals* 9, no. 11 (2019): 862, https://doi.org/10.3390/ani9110862.

horses' facial expressions: Jen Wathan et al., "EquiFACS: The Equine Facial Action Coding System," *PLOS One* 10, no. 9 (August 2015): e0131738, https://doi.org/10.1371/journal.pone.0131738.

ninety-second rule of emotion: Bryan E. Robinson, "The 90-Second Rule That Builds Self-Control," *Psychology Today*, April 26, 2020, https://www.psychologytoday.com/ca/blog/the-right-mindset/202004/the-90-second-rule-builds-self-control.

rats overwhelmingly chose the pedal: G. D. Jensen, "Preference for Bar Pressing over 'Freeloading' as a Function of Number of Rewarded Presses," *Journal of Experimental Psychology* 65, no. 5 (1963): 451–54, https://psycnet.apa.org/doi/10.1037/h0049174.

cats only species that didn't: Mikel M. Delgado, Brandon Sang Gyu Han, and Melissa J. Bain, "Domestic Cats (*Felis catus*) Prefer Freely Available Food over Food That Requires Effort," *Animal Cognition* 25 (2022): 95–102, https://doi.org/10.1007/s10071-021-01530-3.

a surprising 40.9 percent of cats: Mikel M. Delgado et al., "Making Fetch Happen: Prevalence and Characteristics of Fetching Behavior in Owned Domestic Cats (*Felis catus*) and Dogs (*Canis familiaris*)," *PLOS One* 19, no. 9 (September 2024): e0309068, https://doi.org/10.1371/journal.pone.0309068.

Mikel Delgado quotes: Dr. Mikel M. Delgado, in conversation with the author, April 2024.

Turid Rugaas quotes: Turid Rugaas, in conversation with the author, April 2024.

a 2023 study on zebra finches: Katharina Riebel and Naomi E. Langmore, "Birdsong: Not All Contest but Also Cooperation?," *Current Biology* 33, no. 2 (January 2023): 67–69, https://doi.org/10.1016/j.cub.2022.12.015.

female birds sing too: Karan J. Odom et al., "Female Song Is Ancestral and Widespread in Songbirds," *Nature Communications* 5 (2014): 3379, https://doi.org/10.1038/ncomms4379.

CHAPTER 9: TRAINING THE UNTRAINABLE

"All his life he'd been good": Rupert Brooke, *The Collected Poems of Rupert Brooke* (Dodd, Mead, 1931), 181.

Shawna Karrasch quotes: Shawna Karrasch, in conversation with the author, May 2024.

"You cannot know what freedom means": J. A. Baker, *The Peregrine* (William Collins, 2017), 179.

Maasa Nishimuta and Sean Will quotes: Maasa Nishimuta and Sean Will, in conversation with the author, May 2024.

chickens are killed: Max Roser, "How many animals get slaughtered every day?" *Our World in Data*, September 26, 2023, https://ourworldindata.org/how-many-animals-get-slaughtered-every-day.

"acts of kindness": William Wordsworth, *The Complete Poetical Works*, 86.

"Connection before concepts": Sarah Schlote, "Connection Before Concepts: A Comparison of 3 Pressure-Release Methods," Equusoma, Horse-Human Trauma Recovery, March 18, 2020, https://equusoma.com/connection-before-concepts/.

"They do not sweat and whine": Whitman, *Song of Myself* (Dover, 2012), 28.

"When I bestride him": William Shakespeare, *Henry V* (Oxford University Press, 1998), 197.

CHAPTER 10: BACK ON TRACK

"Several of Nature's People": Emily Dickinson, *The Poems of Emily Dickinson* (Little, Brown, 1930), 79.

"characterized by the belief that the parts": "holistic," Dictionary, Google, accessed January 5, 2025, https://www.google.com/search?q=definition+holistic.

wildlife tracker Josh Lane: "Bird Language Basics Free E-Course," Bird Language, accessed February 22, 2025, https://birdlanguage.com/discover /ecourse/.

wild raven calls: Stephan A. Reber et al., "Territorial Raven Pairs Are Sensitive to Structural Changes in Simulated Acoustic Displays of Conspecifics," *Animal Behaviour* 116 (June 2016): 153–62, https://doi.org/10.1016/j.anbehav.2016 .04.005.

pheasants have specialized calls: "Ring-Necked Pheasant," All About Birds, Cornell Lab of Ornithology, accessed February 22, 2025, https://www .allaboutbirds.org/guide/Ring-necked_Pheasant/sounds.

George Bumann quotes: George Bumann, in conversation with the author, June 2024.

Louis Liebenberg quotes: Louis Liebenberg, in conversation with the author, June 2024.

Tamarack Song quotes: Tamarack Song, in conversation with the author, June 2024.

"It is what the sailor holding": Mihaly Csikszentmihalyi, *Flow: The Psychology of Optimal Experience* (HarperCollins, 2008), 3.

"suggest that the psychological experience of loss": Brittany Greene and Jennifer Vonk, "Is Companion Animal Loss Cat-Astrophic? Responses of Domestic Cats to the Loss of Another Companion Animal," *Applied Animal Behaviour Science* 277 (August 2024): 106355, https://doi.org/10.1016/j .applanim.2024.106355.

"Bacteria appear to live lives": Gunnar de Winter, James P. Stratford, and Ben B. Chapman, "Using Bacteria to Study Consistent Variation in Individual Behavior," *Behavioral Ecology* 26, no. 6 (November-December 2015): 1465–69, https://doi.org/10.1093/beheco/arv154.

Kalahari trackers were recently able to identify: Tilman Lenssen-Erz et al., "Animal Tracks and Human Footprints in Prehistoric Hunter-Gatherer Rock Art of the Doro! nawas Mountains (Namibia), Analysed by Present-Day Indigenous Tracking Experts," *PLOS One* 18, no. 9 (September 2023): e0289560, https://doi.org/10.1371/journal.pone.0289560.

"he was not an elk": Rane Willerslev, *Soul Hunters: Hunting, Animism and Personhood among the Siberian Yukaghirs* (University of California Press, 2007), 1.

CHAPTER 11: THE MEDIUM IS THE MESSAGE

"does not keep pace": Thoreau, *Walden*, 323.

"what is any revelation": William J. Long, *How Animals Talk* (Dover, 2009), 26.

"Forces or events": *Cambridge Dictionary*, s.v. "supernatural," accessed January 5, 2025, https://dictionary.cambridge.org/dictionary/english/supernatural.

author Richard Webster says: Richard Webster, *Your Psychic Pet* (Castle Books, 2003).

Penelope Smith quotes: Penelope Smith, in conversation with the author, July 2024.

"he was not an elk": Rane Willerslev, *Soul Hunters: Hunting, Animism and Personhood among the Siberian Yukaghirs* (University of California Press, 2007), 1.

Diana Delmonte quotes: Diana Delmonte, in conversation with the author, July 2024.

"spooky action at a distance": J. S. Bell, *Speakable and Unspeakable in Quantum Mechanics* (Cambridge University Press, 1987), 143.

a 1997 experiment by Swiss physicist Nicolas Gisin: W. Tittel et al., "Long-Distance Bell-Type Tests Using Energy-Time Entangled Photons," *Physical Review A* 59, no. 6 (June 1999): 4150, https://doi.org/10.1103/PhysRevA.59.4150.

Rupert Sheldrake also uses quantum entanglement: "The Extended Mind: Recent Experimental Evidence," Rupert Sheldrake, accessed January 5, 2025, https://www.sheldrake.org/videos/the-extended-mind-recent-experimental-evidence.

he also tested Aimee Morgana and N'kisi: Rupert Sheldrake, "Testing a Language-Using Parrot for Telepathy," *Journal of Scientific Exploration* 17, no. 4 (2003): 601–16, https://www.sheldrake.org/files/pdfs/papers/Testing-a -Language-Using-Parrot-for-Telepathy.pdf.

Aimee Morgana quotes: Aimee Morgana, email correspondence with the author, July 2024.

"You should talk": Kathy Price, in conversation with the author, July 2024.

"stiff with magic": Roger Deakin, *Waterlog, A Swimmer's Journey Through Britain* (Tin House, 2021), 100.

"Seeing a luna moth": Cheyenne Main, "Spiritual Symbolism of the Luna Moth: Plus, What to Do If You See One," wikiHow, August 12, 2023, https:// www.wikihow.com/Luna-Moth-Meaning.

"signify new beginnings": "The Luna Moth & the Full Moon, The Spiritual Significance of Signs That Appear in Our Lives," Inner Wild Woman Studio, May 23, 2024, https://www.innerwildwomanstudio.com/blog/the-luna-moth -the-full-moon-the-spiritual-significance-of-signs-that-appear-in-our-lives.

"are symbols of transformation": Annabelle Reyes, "What Does a Moth Symbolize?," wikiHow, September 28, 2024, https://www.wikihow.com/Moth -Symbolism.

"I crossed the threshold": Święch, "Paradise Called Dziedzinka."

"Every little pine needle": Thoreau, *Walden*, 130.

"IIC presents as a detailed": "Intuitive Interspecies Communication (IIC): What Is It?," Research, Intuitive Interspecies Communication, University of Saskatchewan, accessed February 22, 2025, https://researchers.usask.ca/mj -barrett/research.php.

"In simple terms": Brian E. Neubauer, Catherine T. Witkop, and Lara Varpio, "How Phenomenology Can Help Us Learn from the Experiences of Others," *Perspectives on Medical Education* 8, no. 2 (April 2019): 90–97, https://doi.org /10.1007/s40037-019-0509-2.

"My ancestors could": "About Me," Sydney Kuppenbender, accessed February 22, 2025, https://www.sydneykuppenbender.ca/about-me.

tricksters of First Nations and Native American stories: "Métis Mythology and Folklore: Mythological Figures," Virtual Museum of Métis History and Culture, Gabriel Dumont Institute of Native Studies and Applied Research, accessed February 22, 2025, https://www.metismuseum.ca/media/document .php/13465.Metis%20Mythology%20and%20Folklore.pdf.

"Métis stories are not 'make-believe'": "Oral Tradition," *Indigenous Peoples Atlas of Canada*, accessed February 22, 2025, https://indigenouspeoplesatlasofcanada .ca/article/oral-tradition/.

M. J. Barrett quotes: Dr. M. J. Barrett, in conversation with the author, August 2024.

"challenges deeply inscribed Cartesian dualisms": M. J. Barrett et al., "Speaking with Other Animals Through Intuitive Interspecies Communication: Towards Cognitive and Interspecies Justice," in *A Research Agenda for Animal Geographies*, ed. Alice Hovorka, Sandra McCubbin, and Lauren Van Patter (Edward Elgar, 2021), 158.

"interspecies humility": Barrett et al., "Speaking with Other Animals."

"We see things not as they are but as we are": George T. W. Patrick, "The Psychology of Prejudice," *Popular Science Monthly* 36 (1890): 633, https://en .wikisource.org/wiki/Popular_Science_Monthly/Volume_36/March_1890/The _Psychology_of_Prejudice.

Oscar the cat: David M. Dosa, "A Day in the Life of Oscar the Cat," *New England Journal of Medicine* 357, no. 4 (July 2007): 328–29, https://doi.org/10 .1056/NEJMp078108.

dogs prefer aligning themselves: Vlastimil Hart et al., "Dogs Are Sensitive to Small Variations of the Earth's Magnetic Shield," *Frontiers in Zoology* 10 (December 2013): 80, https://doi.org/10.1186/1742-9994-10-80.

CHAPTER 12: AN UNFAMILIAR FEELING

"silence any song": Robert Frost, *Complete Poems of Robert Frost* (Holt, Rinehart and Winston, 1964), 316.

"The majority of pigs": "Rearing Kunekune Pigs for Meat," British Kunekune

Pig Society, accessed February 22, 2025, https://www.britishkunekunesociety.org.uk/raising-for-meat/.

"marked by compassion, sympathy": *Merriam-Webster Dictionary*, s.v. "humane," accessed January 5, 2025, https://www.merriam-webster.com/dictionary/humane.

about 150 species are vanishing from the planet every day: "Indigenous and Local Communities—the Human Face of Climate Change," Convention on Biological Diversity, May 22, 2007, https://www.cbd.int/doc/speech/2007/sp-2007-05-22-unpfii-en.pdf.

"We patronize them for their incompleteness": Henry Beston, *The Outermost House: A Year of Life on the Great Beach of Cape Cod* (Doubleday, Doran, 1928), 2.

American veterinarians are three to five times: Katherine Compitus, "Silent Suffering: The High Rate of Suicide in Veterinarians," *Psychology Today*, November 17, 2023, https://www.psychologytoday.com/ca/blog/zooeyia/202311/silent-suffering-the-high-rate-of-suicide-in-veterinarians.

the toll of "compassion fatigue": Mary Lee Jensvold, "A Preliminary Assessment of Compassion Fatigue in Chimpanzee Caretakers," *Animals* 12, no. 24 (December 2022): 3506, https://doi.org/10.3390/ani12243506.

"I finished my studies in biology": Święch, "Paradise Called Dziedzinka."

Manjiri Latey quotes: Manjiri Latey, in conversation with the author, August 2024.

a podcast for kids: "Where Does the Word 'Word' Come From?," August 1, 2022, in *Happy Podcast for Kids*, Anorak, https://anorakmagazine.com/pages/happy-podcast-for-kids?srsltid=AfmBOopOPLQfQhmvfn88qk2k4Fc_gGO.

"Recover the gifts": "Interspecies Consciousness," Generation of Harmony, accessed January 5, 2025, https://www.generateharmony.com/interspecies-consciousness.

Kerri Lake quotes: Kerri Lake, in conversation with the author, August 2024.

"My heart in hiding": Gerard Manley Hopkins, *Poems of Gerard Manley Hopkins* (Humphrey Milford, 1918), 29.

"The Poetry of earth": John Keats, Complete Poems (Harvard University Press, 1982), 54.

"There is sorror enough": Rudyard Kipling, *Rudyard Kipling's Verse Inclusive Edition 1885–1918* (Doubleday, Page, 1922), 656.

"Happy we who can bask": Henry David Thoreau, *A Week on the Concord and Merrimack Rivers* (Houghton, Mifflin, 1883), 74.

EPILOGUE: I SAY GOODBYE, AND YOU SAY HELLO

"pick out the gravel": John Keats, *Selected Letters*, 35.

"infinite and unaccountable friendliness": Thoreau, *Walden*, 130.

"illusion of separateness": David Loy, "Awakening from the Illusion of Our Separateness," Huffpost, 11 October 2011, https://www.huffpost.com/entry /awakening-from-the-illusion-of-our-separateness_b_988590.

"A human being is part of the whole": Bryce Haymond, "Einstein's Misquote on the Illusion of Feeling Separate from the Whole," Thy Mind, O Human, accessed January 5, 2025, https://www.thymindoman.com/einsteins-misquote -on-the-illusion-of-feeling-separate-from-the-whole/.

the Jain Buddhist monks: Ron Cherry and Hardev S. Sandhu, "Insects in the Religions of India," *American Entomologist* 59, no. 4 (October 2013): 202, http:// dx.doi.org/10.1093/ae/59.4.200.

"Nothing is especially wonderful": Long, *How Animals Talk*, 34.

About the Author

Amelia Thomas is a Cambridge University–educated British author, naturalist, and journalist who lives with her husband and children on a farm in Canada. A former travel writer, she has written and contributed to more than a dozen books for Lonely Planet and has worked for newspapers and magazines worldwide. Her first nonfiction book, *The Zoo on the Road to Nablus*, tells the true story of the last Palestinian zoo.

www.ingramcontent.com/pod-product-compliance
Lightning Source LLC
Chambersburg PA
CBHW021613270326
41931CB00008B/683